SpringerBriefs in Physics

SpringerBriefs in Physics are a series of slim high-quality publications encompassing the entire spectrum of physics. Manuscripts for SpringerBriefs in Physics will be evaluated by Springer and by members of the Editorial Board. Proposals and other communication should be sent to your Publishing Editors at Springer.

Featuring compact volumes of 50 to 125 pages (approximately 20,000–45,000 words), Briefs are shorter than a conventional book but longer than a journal article. Thus, Briefs serve as timely, concise tools for students, researchers, and professionals.

Typical texts for publication might include:

- A snapshot review of the current state of a hot or emerging field
- A concise introduction to core concepts that students must understand in order to make independent contributions
- An extended research report giving more details and discussion than is possible in a conventional journal article
- A manual describing underlying principles and best practices for an experimental technique
- An essay exploring new ideas within physics, related philosophical issues, or broader topics such as science and society

Briefs allow authors to present their ideas and readers to absorb them with minimal time investment. Briefs will be published as part of Springer's eBook collection, with millions of users worldwide. In addition, they will be available, just like other books, for individual print and electronic purchase. Briefs are characterized by fast, global electronic dissemination, straightforward publishing agreements, easy-to-use manuscript preparation and formatting guidelines, and expedited production schedules. We aim for publication 8–12 weeks after acceptance.

Stéphane Peigné

Color in QCD

An Introduction Featuring the Birdtrack
Pictorial Technique

 Springer

Stéphane Peigné
Nantes, France

ISSN 2191-5423   ISSN 2191-5431 (electronic)
SpringerBriefs in Physics
ISBN 978-3-031-53680-9   ISBN 978-3-031-53681-6 (eBook)
https://doi.org/10.1007/978-3-031-53681-6

This Springer imprint is published by the registered company Springer Nature Switzerland AG
The registered company address is: Gewerbestrasse 11, 6330 Cham, Switzerland

Paper in this product is recyclable.

Preface

Quarks and gluons, the elementary particles that make up nucleons (protons and neutrons) and, more generally, all hadrons, interact via the strong interaction described by quantum chromodynamics (QCD). This manual focuses on the *color* of quarks and gluons, the quantum number responsible for their mutual strong interactions, and provides an elementary introduction to the birdtrack technique, which is a powerful tool for addressing the color structure of QCD in a pictorial way, drawing color graphs named *birdtracks* rather than writing mathematical symbols carrying color indices. The birdtrack technique shows how quark and gluon colors are combined and mixed in QCD. We will learn the basic rules, discuss color conservation and infinitesimal color rotations, learn how to project on the color states of systems of quarks, antiquarks, and gluons (called partons), derive their Casimir charges... and at the same time learn a little bit of representation theory.

This manual is primarily intended for particle physics students, graduates, and researchers working in the field of QCD, but may also be of interest to students of mathematics, as an illustration of the use of the birdtrack technique in representation theory. It does not require any specific prerequisite (except perhaps a few notions of linear algebra, which will be recalled anyway) and includes very simple exercises to keep the focus and learn by doing. Doing these exercises is mandatory to avoid superficial reading and to learn how to use birdtracks in practice.

We will consider $SU(N)$ with $N \geq 3$ as the symmetry group of QCD, i.e., N quark colors, $N = 3$ corresponding to real-world QCD. Indeed, for general N the birdtrack technique is not more complicated than for $N = 3$. On the contrary, working with a general parameter N for the number of quark colors usually brings a deeper understanding than working with fixed $N = 3$. The birdtrack technique can be used as a handy tool in virtually any $SU(N)$ color calculation, whether it is evaluating simple color factors or addressing complex color structures that may otherwise seem out of reach.

This introductory manual should provide a basic toolbox for drawing birdtracks, which I hope will be useful for acquiring more in-depth knowledge to tackle advanced problems on the subject.

The content of Chaps. 1–4 is that of the lectures given at the *6th Chilean School of High Energy Physics* (Valparaiso, January 2023). These lectures inspired this manual, and I'd like to thank the organizers of the Chilean school for their invitation, and all the participants for their enthusiastic involvement.

May the present manual give you a taste of the unspeakable satisfaction that can come from drawing birdtracks!

Nantes, France   Stéphane Peigné
December 2023

Contents

Chapter 1
Pedestrian Introduction

Abstract We define the elementary bricks from which all color graphs named *bird-tracks* can be constructed, and derive pictorial expressions of the Fierz identity and the SU(N) Lie algebra. Using these identities, we obtain once and for all a set of simple pictorial rules, which will be used repeatedly throughout this brief.

1.1 The Basic Lego Bricks

The color structure of QCD can be expressed in terms of three elementary bricks, or "legos". Using these legos, the color structure of any QCD problem can in principle be worked out. The legos are defined and represented pictorially as[1]

$$i \longrightarrow j \atop a \quad = (T^a)^j_{\ i} \; ; \qquad i \longleftarrow j \atop a \quad = -(T^a)^i_{\ j} \; ; \qquad c \longleftarrow b \atop a \quad = -if_{abc} \; , \quad (1.1)$$

where a plain line with an arrow represents a quark (which will be denoted by q) or an antiquark ($\bar{q}$), and a curly line a gluon (g).

As mentioned in the preface, we consider the SU(N) symmetry group for $N \geq 3$. Quark and antiquark color indices denoted by $i, j \ldots$ thus vary from 1 to N, and gluon indices denoted by $a, b \ldots$ from 1 to $N^2 - 1$. The matrices T^a (with $a = 1 \ldots N^2 - 1$) are $N \times N$ Hermitian matrices $((T^a)^\dagger \equiv {}^t(T^a)^* = T^a)$, of zero trace,

$$\text{Tr } T^a = 0 \; , \tag{1.2}$$

and normalized so that

$$\text{Tr }(T^a T^b) = \frac{1}{2}\delta_{ab} \; . \tag{1.3}$$

---

[1] In this manual, color graphs are drawn with the TikZ LATEX package (TikZ version 3.1.2).

© The Author(s), under exclusive license to Springer Nature Switzerland AG 2024
S. Peigné, *Color in QCD*, SpringerBriefs in Physics,
https://doi.org/10.1007/978-3-031-53681-6_1

The f_{abc}'s are the SU(N) structure constants defining the Lie algebra of SU(N),

$$\left[T^a, T^b\right] = i f_{abc} T^c , \tag{1.4}$$

from which one can easily show that f_{abc} is totally antisymmetric in a, b, c.

To memorize the three legos defined in (1.1), one may view them as representing respectively quark, antiquark, and gluon scattering (the time arrow going from left to right) off an external gluon field (carrying the color index a) coupled *from below*, and remember that each lego corresponds to the SU(N) *generator* of index a in the representation of the scattered parton, namely, T^a for a quark, $-T^a$ for an antiquark, and t^a for a gluon, where the $(N^2 - 1) \times (N^2 - 1)$ matrices t^a are defined by

$$(t^a)_{bc} = -i f_{abc} . \tag{1.5}$$

Note however that the meaning of a generator (which will be recalled in Chap. 2) is not really useful at this stage.

An important feature of the rules (1.1) is that each lego is antisymmetric under the exchange of two lines. For instance, by exchanging any two lines of the first lego and rotating if needed the resulting graph in the mind's eye, one obtains a diagram looking like the second lego (antiquark scattering off a gluon coupled from below), which thus gets a minus sign. Let us also remark that in (1.1), the color index of an outgoing quark (or incoming antiquark) is written by convention as an upper index, and that of an outgoing antiquark (or incoming quark) as a lower index. The usefulness of this convention will become clear below.

In addition to (1.1), we introduce the pictorial notation for Kronecker's of color indices:

$$i \longrightarrow j = \delta^j_i \quad ; \qquad a \mathrm{\rightsquigarrow} b = \delta_{ab} . \tag{1.6}$$

For those who have not yet had the chance to get acquainted with QCD or even Feynman diagrams, let us mention that in a "lego construction" built from the basic legos, the color index of an internal line is summed over all its possible values. Together with Einstein summation convention, according to which an index appearing twice in an expression is implicitly summed, we thus have, for an internal gluon line:

$$= (T^a)^j_i (-T^a)^l_k = \qquad = \delta_{ab} \qquad = \tag{1.7}$$

and similarly, for an internal quark line:

$$\text{(1.8)}$$

Obviously, in pictorial notation summing over a repeated index amounts to connecting lines. The convention that a line with an arrow pointing out of (into) a vertex carries an upper (lower) index implies that for quarks and antiquarks, a repeated index always appears in both upper and lower positions. In other words, this convention simply serves to distinguish quarks from antiquarks, and is consistent with the connection of quark lines respecting the direction of the arrow.

In what follows, everything will be deduced step by step, using only (1.1) and (1.6) as building blocks. The graphs resulting from the combination of these legos will be called "color graphs" or *birdtracks*, using the terminology of the reference book on this pictorial technique, Cvitanović (2008). Introductions to the birdtrack technique applied to SU(N) can be found in Refs. Dokshitzer (1995), Keppeler (2017), and in Peigné (2023), which includes (with the exception of Sect. 4.2.2). Chaps. 1–4 of this manual.

1.2 First Trivial Rules

The simplest way color appears in QCD is in the form of numbers called *color factors*. For example, in the calculation of the $qq \rightarrow qqg$ partonic cross section, one of the contributions is proportional to

$$\text{(1.9)}$$

where each factor is a *matrix element* in color space (i.e., a color graph with specified external indices), and every initial and final color index is implicitly summed. To calculate such color factors in a pictorial way, we first need the following rule:

> The complex conjugate of a color matrix element is obtained pictorially by taking the mirror image of the associated graph and reversing the arrows of quark and antiquark lines.

Exercise 1.1 Check that this rule indeed holds for each of the three legos (1.1), and then show that it must be true for any matrix element built from these legos.

Applying the complex conjugation rule to the second factor of (1.9) and then summing over repeated indices by connecting lines, the color factor (1.9) becomes

$$(1.10)$$

In order to evaluate this type of "connected graph", we will need simple pictorial rules. The most trivial ones read

$= $ $_i^i = \delta^i{}_i = N$, $\qquad (1.11)$

$=$ $_a^a = \delta_{aa} = N^2 - 1$, $\qquad (1.12)$

$= (T^a)^i{}_i = 0$, $\qquad (1.13)$

$= (T^a)^j{}_i (T^b)^i{}_j = \dfrac{1}{2}$, $\qquad (1.14)$

where (1.13) and (1.14) follow from Eqs. (1.2) and (1.3). In the following sections, more interesting rules are obtained using pictorial representations of the Fierz identity and of the Lie algebra.

1.3 Fierz Identity

In index notation, the Fierz identity reads

$$\delta^i{}_j \, \delta^l{}_k = 2 \, (T^a)^i{}_k \, (T^a)^l{}_j + \frac{1}{N} \, \delta^i{}_k \, \delta^l{}_j \ , \qquad (1.15)$$

which corresponds pictorially to

$\qquad (1.16)$

For more comfort, we may rewrite the latter equation by removing external indices:

$$\text{―} = 2 \left[\text{≻mm≺} + \frac{1}{N} \text{][} \right] . \tag{1.17}$$

It is important to note that in doing so, the mathematical meaning of the pictorial equation is changed: a color graph represents a *color matrix element* when external indices are specified, or the corresponding *linear map* (between the vector spaces spanned by the objects carrying the initial and final indices) when external indices are removed. For instance,

$$\begin{array}{c} i \\ j \\ a \end{array} \left(A \right) \begin{array}{c} k \\ l \end{array} \equiv A^{kl}{}_{ija} \tag{1.18}$$

are the matrix elements of the operator

$$\left(A \right) \equiv A . \tag{1.19}$$

Note that the pictorial rule for complex conjugation of matrix elements corresponds to *Hermitian conjugation* at the operator level. Indeed, the complex conjugate of (1.18) reads (using $A^* = {}^t A^\dagger$)

$$(A^{kl}{}_{ija})^* = (A^\dagger)^{ija}{}_{kl} = \begin{array}{c} k \\ l \end{array} \left(A^\dagger \right) \begin{array}{c} i \\ j \\ a \end{array} , \tag{1.20}$$

but according to the complex conjugation rule, this must coincide with the graph obtained by taking the mirror image of (1.18) and reversing arrows. Thus, that transformation applied to the operator A must give $A^\dagger$:

> The Hermitian conjugate of an operator is obtained pictorially by taking the mirror image of the corresponding graph and reversing quark and antiquark arrows.

The Fierz identity (1.17) is thus a relation between linear maps of the vector space $\mathcal{V} \otimes \overline{\mathcal{V}}$ to itself, where $\mathcal{V} \equiv \{q^i\}$ and $\overline{\mathcal{V}} \equiv \{q_i\}$ are the quark and antiquark vector spaces, respectively. In particular, the l.h.s. of (1.17) is the identity operator:

$$\text{―} = \mathbb{1}_{\mathcal{V} \otimes \overline{\mathcal{V}}} . \tag{1.21}$$

Exercise 1.2 Prove the Fierz identity (1.17) by using the following linear algebra result: if $\mathcal{E}$ is a vector space of dimension n, the operators p_i ($i = 1 \ldots m$) are projectors ($p_i^2 = p_i$) such that $p_i p_j = 0$ for $i \neq j$, and $\sum_{i=1}^m \text{rank}(p_i) = n$ (where rank$(f) \equiv \dim[\text{img}(f)]$), then $\sum_{i=1}^m p_i = \mathbb{1}_{\mathcal{E}}$.
Hint: Use the fact that the rank of a projector is equal to its trace, and that the trace is simply obtained pictorially by connecting the initial and final lines carrying the same type of indices. For instance, the rank of the identity projector (1.21) can be written as:

$$\text{rank}(\mathbb{1}_{V \otimes \overline{V}}) = \text{Tr}\,(\mathbb{1}_{V \otimes \overline{V}}) = \text{Tr}\,(\;\overrightarrow{}\;) = \;\bigcirc\!\!\!\!\!\bigcirc\; = N^2, \qquad (1.22)$$

which coincides as it should with the dimension of $V \otimes \overline{V}$.

Let us now use the Fierz identity to derive simple pictorial rules.

Exercise 1.3 Multiply (1.16) by $i \longrightarrow j = \delta^j{}_i$ and sum over repeated indices to show that

$$\overrightarrow{} = C_{\text{F}} \longrightarrow , \qquad (1.23)$$

$$C_{\text{F}} \equiv \frac{N^2 - 1}{2N} . \qquad (1.24)$$

Note that (1.23) can be written as $T^a T^a = C_{\text{F}} \mathbb{1}_V$, where $T^a T^a$ is called the Casimir operator in the fundamental (quark) representation, and C_{F} the (squared) quark color charge or simply "quark Casimir". (In any SU(N) irreducible representation R, the Casimir operator $T^a(R)T^a(R)$ is proportional to the identity in that representation, as a consequence of Schur's lemma, see Chap. 3.)

Exercise 1.4 Use the Fierz identity to obtain:

$$\overrightarrow{} = -\frac{1}{2N} \overrightarrow{} . \qquad (1.25)$$

1.4   Lie Algebra

1.4.1   *Lie Algebra in the Fundamental Representation*

The matrices T^a (called the *generators* of SU(N) in the fundamental representation, see Chap. 2) satisfy the Lie algebra (1.4), which can be expressed pictorially as

$$\overrightarrow{} = \overrightarrow{} + \overrightarrow{} . \qquad (1.26)$$

Exercise 1.5 Check it!

The identity (1.26) is the first example of *color conservation* that we encounter. (See Chap. 2 for a general discussion of color conservation.) We will now use (1.26) to find new simple rules.

Exercise 1.6 Using *only* the pictorial rules derived so far, show that

$$= N \quad . \tag{1.27}$$

Hint: To start, trade a gluon for a quark-antiquark pair by using (1.14).

Exercise 1.7 Obtain in two different ways the rule:

$$= \frac{N}{2} \quad . \tag{1.28}$$

Hint: Multiply (1.26) *either to the left by* , *or to the right by* , *and sum over appropriate indices.*

1.4.2 Lie Algebra in the Adjoint Representation

A nice identity is the so-called Jacobi identity,

$$= + \quad , \tag{1.29}$$

which is another manifestation of color conservation.

Exercise 1.8 Prove the Jacobi identity in two ways:

1. Verify that $[T^a, [T^b, T^c]] + [T^b, [T^c, T^a]] + [T^c, [T^a, T^b]] = 0$ to deduce the relation $f_{abe} f_{cde} + f_{bce} f_{ade} + f_{cae} f_{bde} = 0$, and check that the latter is equivalent to (1.29).
2. The previous proof indicates that (1.29) is a direct consequence of the Lie algebra in the fundamental representation (1.4). Thus, it should be possible to prove the Jacobi identity (1.29) pictorially, using only (1.26) and the pictorial rules derived so far. Try it!
   Hint: Use a similar start as in Exercise 1.6.

Exercise 1.9 Check that the Jacobi identity is nothing but the expression of the $SU(N)$ Lie algebra in the adjoint representation, namely,

$$\left[t^a, t^b\right] = i f_{abc} \, t^c , \tag{1.30}$$

where the matrices t^a are defined by (1.5).

Anticipating the next chapters, we note that since the t^a matrices satisfy the Lie algebra and also $(t^a)^\dagger = t^a$ and $\mathrm{Tr}\, t^a = 0$, they are the SU(N) generators (see Chap. 2) of the representation of dimension $N^2 - 1$ called the gluon or *adjoint* representation. The operator $t^a t^a$ is thus the Casimir operator in the adjoint representation, and from Schur's lemma (see Chap. 3) we must have $t^a t^a = C_A \mathbb{1}_{\mathcal{A}}$, where $\mathcal{A} \equiv \{g^a\}$ denotes the gluon vector space of dimension $N^2 - 1$, and C_A the "gluon Casimir". In fact, we already found $t^a t^a$, since Eq. (1.27) reads (in index notation):

$$f_{acd}\, f_{bcd} = (t^c)_{ad} (t^c)_{db} = (t^c t^c)_{ab} = N \delta_{ab}, \tag{1.31}$$

implying that $t^c t^c = N\, \mathbb{1}_{\mathcal{A}}$ and that the gluon Casimir is $C_A = N$.

Exercise 1.10 Use the Jacobi identity to obtain the rule:

$$\text{} = \frac{N}{2}\ \text{}. \tag{1.32}$$

1.5 Sum Up

In this first chapter, we have obtained the following set of simple pictorial rules (recall that C_F is defined in (1.24)):

$$\tag{1.33}$$

These rules will prove very useful in the next chapters. When one of the above loop diagrams appears as a sub-diagram in a color graph, it can be replaced by its simpler r.h.s. expression given in (1.33).

Exercise 1.11 Using the rules (1.33), calculate the color factor (1.10).

Exercise 1.12 Evaluate the color graph . (Borrowed from Dokshitzer (1995).)

Chapter 2
Color Conservation, Color Rotations, and SU(N) Irreducible Representations

Abstract After defining color conservation, we discuss SU(N) transformations or "color rotations" and their action on quark, antiquark and gluon color indices, allowing to introduce the notion of SU(N) representation. Considering infinitesimal color rotations, we observe that color conservation is equivalent to the SU(N) invariance of color singlet parton systems. Finally, we show how the *irreducible representations* (irreps) of a multi-parton system emerge pictorially.

2.1 Color Conservation Pictorially

In Chap. 1, we wrote the pictorial form of the Lie algebra, both in the fundamental and adjoint representation (see (1.26) and (1.29)):

$$ \tag{2.1} $$

$$ \tag{2.2} $$

The latter equations can also be written by stretching the incoming gluon line to the final state (or the outgoing parton lines to the initial state). For instance, stretching the incoming gluon in (2.1) to the final state, one obtains (due to the antisymmetry of the lego bricks (1.1) under the exchange of two lines)

$$ 0 = \tag{2.3} $$

Thus, for each lego the sum of gluon attachments to parton lines "before" and "after" the interaction vertex gives the same result, and this holds independently of the way the lego is represented in a time-ordered picture: either as $1 \to 2$ splitting (as in (2.1) and (2.2)), or $2 \to 1$, or $3 \to 0$, or $0 \to 3$ as in (2.3).

© The Author(s), under exclusive license to Springer Nature Switzerland AG 2024
S. Peigné, *Color in QCD*, SpringerBriefs in Physics,
https://doi.org/10.1007/978-3-031-53681-6_2

This trivially generalizes to any operator constructed from the lego bricks, leading to the pictorial representation of *color conservation*:

$$(2.4)$$

where an ellipse crossed by a set of parton lines denotes the sum of all attachments of the "scattering gluon" (drawn below the ellipse in Eq. 2.4) to those lines.

Exercise 2.1 Although quite trivial, give a convincing proof of (2.4).

2.2 Color Rotations

2.2.1 SU(N) Transformations

The special unitary group $SU(N)$ is the Lie group of $N \times N$ unitary matrices with unit determinant,

$$U \in SU(N) \iff UU^\dagger = \mathbb{1} \text{ and } \det U = 1. \tag{2.5}$$

Any element of $SU(N)$ can be parametrized by

$$U(\alpha) = e^{i\alpha^a T^a}, \tag{2.6}$$

where the matrices T^a $(a = 1 \ldots N^2 - 1)$ are the Hermitian matrices introduced in Chap. 1 (hence the name of $SU(N)$ *generators* for those matrices). The $N^2 - 1$ real parameters α^a may be viewed as the "angles" of the "color rotation" $U(\alpha)$.

Exercise 2.2 Verify that the matrix (2.6) indeed belongs to $SU(N)$. (In fact, the exponential parametrization (2.6) generates *all* elements of $SU(N)$, see mathematics textbooks for a proof.)

By construction, the QCD Lagrangian is invariant under $SU(N)$ color rotations (Collins 2011). In order to address the color structure of QCD (in particular, to determine the *irreps* and associated *multiplets* of a parton system, see Sect. 2.3), we first describe the action of color rotations on quark, antiquark and gluon "color coordinates".

2.2.2 Color Rotations of Quark and Antiquark Coordinates

Let us start with quarks and antiquarks. Under a given color rotation $U(\alpha) \in SU(N)$, the quark coordinates q^i (denoted by an *upper* index according to our initial convention, see Sect. 1.1) transform as

$$q'^i = U^i{}_j \, q^j \quad \Leftrightarrow \quad q' = U \, q \, . \tag{2.7}$$

When we restrict ourselves to the color degree of freedom (and ignore the spin of quarks, antiquarks and gluons), as we do in this manual, antiquark coordinates are simply obtained from quark coordinates by complex conjugation. In the same color rotation $U(\alpha)$, antiquark coordinates thus transform as

$$(q^{*\prime})^i = (U^*)^i{}_j \, (q^*)^j \quad \Leftrightarrow \quad q^{*\prime} = U^* \, q^* \, . \tag{2.8}$$

A standard convention is to denote complex conjugation by moving quark and antiquark indices up and down, namely,

$$(q^*)^i \equiv q_i \; ; \quad (U^*)^i{}_j \equiv U_i{}^j \, , \tag{2.9}$$

a convention that we have implicitly used from the beginning by assigning *lower* color indices to antiquarks, see Sect. 1.1. The transformation (2.8) of antiquark coordinates is then written as

$$q'_i = U_i{}^j \, q_j \, . \tag{2.10}$$

To complement the above convention, any quantity transforming as quark (antiquark) coordinates is assigned an upper (lower) index. We readily verify that a product of the form $A^i B_i$ (implicitly summed over i) is $SU(N)$ invariant. Indeed:

$$(A^i B_i)' = U^i{}_j \, U_i{}^k \, A^j B_k = A^i B_i \tag{2.11}$$

follows directly from

$$U^i{}_j \, U_i{}^k = U^i{}_j (U^*)^i{}_k = U^i{}_j (U^\dagger)^k{}_i = (U^\dagger U)^k{}_j = \delta^k{}_j \; . \tag{2.12}$$

Under two successive color rotations of angles α^a and β^b, quark coordinates transform as

$$q \xrightarrow{\alpha} U(\alpha) \, q \xrightarrow{\beta} U(\beta) U(\alpha) \, q = U(\gamma(\alpha, \beta)) \, q \, . \tag{2.13}$$

Here, we've used the fact that $SU(N)$ is a group, so the product $U(\beta)U(\alpha)$ must coincide with an element of $SU(N)$ of angles γ^c, the latter being fully determined by α^a and β^b.

Exercise 2.3 Here is a standard exercise that should be done once (or twice) in a lifetime. Let us recall the Baker-Campbell-Hausdorff formula for the product of two exponentials of matrices:

$$e^X e^Y = e^{X+Y+\frac{1}{2}[X,Y]+\frac{1}{12}([X,[X,Y]]-[Y,[X,Y]])+\cdots} , \qquad (2.14)$$

where the dots stand for higher-order terms in X and Y (all being nested commutators of X and Y). Using (2.14), show that the angles γ^a defined by (2.13) are given by $\gamma^a(\alpha,\beta) = \alpha^a + \beta^a + \frac{1}{2}f_{abc}\alpha^b\beta^c + \ldots$, and find the next term in the series.

This exercise illustrates that the structure of SU(N) (with respect to the multiplication law) is fully determined by the SU(N) Lie algebra (1.4).

Let's end this section by a bit of terminology:

The set of $N \times N$ matrices $U(\alpha) = e^{i\alpha^a T^a}$ (i.e. the SU(N) group itself) acting on the quark is called the *fundamental representation* of SU(N). The antiquark transforms under the complex conjugate representation, defined by the set of $N \times N$ matrices $U(\alpha)^* = e^{-i\alpha^a(T^a)^*}$. The SU($N$) quark and antiquark representations have dimension N (the dimension of the objects on which they act) but are conventionally named **3** and $\bar{\mathbf{3}}$, respectively, from the value of their dimension in the case $N = 3$. The SU(N) generators in these representations are the matrices $T^a(\mathbf{3}) = T^a$ and $T^a(\bar{\mathbf{3}}) = -(T^a)^*$.

Note that while the fundamental representation **3** and its complex conjugate $\bar{\mathbf{3}}$ have the same dimension N, they are not equivalent (for $N \geq 3$), i.e., there is no change of basis relating $U(\alpha)$ and $U(\alpha)^*$ for all α. For $N \geq 3$ these representations therefore describe the transformations of intrinsically different objects.

2.2.3 Color Rotations of Gluon Coordinates

The gluon vector space $\mathcal{A} \equiv \{g^a\}$ is of dimension $K_A \equiv N^2 - 1$. How should the gluon coordinates g^a transform under a color rotation of angles α^a, knowing that the N quark coordinates transform with the matrix $U(\alpha)$? For each $U(\alpha)$ acting in quark space, we must find a corresponding $K_A \times K_A$ matrix $\tilde{U}(\alpha)$ acting in gluon space, in such a way that the "representation" $U(\alpha) \to \tilde{U}(\alpha)$ preserves the group structure. Indeed, in the successive rotations of angles α and β, the gluon coordinates become

$$g \xrightarrow{\alpha} \tilde{U}(\alpha) g \xrightarrow{\beta} \tilde{U}(\beta)\tilde{U}(\alpha) g , \qquad (2.15)$$

but for consistency with (2.13), the same result should be obtained by a single rotation of angles $\gamma(\alpha,\beta)$, represented by $\tilde{U}(\gamma(\alpha,\beta))$ when acting in gluon space. We must therefore have the relation

$$\tilde{U}(\beta)\tilde{U}(\alpha) = \tilde{U}(\gamma(\alpha, \beta)), \tag{2.16}$$

with the same function $\gamma(\alpha, \beta)$ as derived in Exercise 2.3.

It is clear that (2.16) will be satisfied by the matrices

$$\tilde{U}(\alpha) = e^{i\alpha^a \tilde{T}^a} \tag{2.17}$$

if one can find a set of $K_A \times K_A$ matrices $\tilde{T}^a$ ($a = 1 \ldots N^2 - 1$) having the same Lie algebra as the T^a's, namely,

$$\left[\tilde{T}^a, \tilde{T}^b\right] = i f_{abc} \tilde{T}^c. \tag{2.18}$$

We know that the matrices t^a defined by (1.5) satisfy these conditions, see (1.30).

The set of matrices $\tilde{U}(\alpha) = e^{i\alpha^a t^a}$ acting on the gluon is called the SU(N) *adjoint representation*, of dimension $K_A = N^2 - 1$, named **8** and of generators $T^a(\mathbf{8}) = t^a$. This representation is real: $\tilde{U}(\alpha)^* = e^{-i\alpha^a (t^a)^*} = e^{i\alpha^a t^a} = \tilde{U}(\alpha)$.

2.2.4   Infinitesimal Color Rotations

Since SU(N) is a Lie group, its structure can be understood by considering only infinitesimal transformations (Georgi 2000). In particular, it is sufficient to consider infinitesimal transformations to highlight SU(N) irreducible representations (see Sect. 2.3).

Let us consider a color rotation of infinitesimal angles $\delta\alpha^a$. According to (2.6) and (2.7), the quark transforms as

$$q'^i = q^i + i\delta\alpha^a (T^a)^i_{\ j} q^j, \tag{2.19}$$

from which the transformation of the antiquark directly follows (take the complex conjugate, and recall that $(T^a)^* = {}^t T^a$):

$$q'_i = q_i - i\delta\alpha^a q_j (T^a)^j_{\ i}. \tag{2.20}$$

Using (2.17), the gluon transforms as:

$$g'^b = g^b + i\delta\alpha^a (t^a)_{bc} g^c. \tag{2.21}$$

The *infinitesimal shifts* of the quark, antiquark and gluon coordinates thus read

$$\delta q^i \equiv q'^i - q^i = i\delta\alpha^a \quad \text{(diagram)}_a^{\; i} \; ,$$

$$\delta q_i \equiv q'_i - q_i = i\delta\alpha^a \quad \text{(diagram)}_a^{\; i} \; , \tag{2.22}$$

$$\delta g^b \equiv g'^b - g^b = i\delta\alpha^a \quad \text{(diagram)}_a^{\; b} \; ,$$

where we introduced the pictorial notation for coordinates:

$$\text{(diagram)}\; j \equiv q^j \; ; \quad \text{(diagram)}\; j \equiv q_j \; ; \quad \text{(diagram)}\; c \equiv g^c \; . \tag{2.23}$$

The basic legos (1.1) are defined as the SU(N) generators in the quark, antiquark and gluon representations, and multiplying by the factor $i\delta\alpha^a$, we obviously get the infinitesimal shifts (2.22).

In a color rotation of angles $\delta\alpha^a$, the infinitesimal shift of parton coordinates is obtained pictorially (up to the factor $i\delta\alpha^a$) by attaching a gluon of color index a from below to the corresponding line.

Let us now rewrite the expression (2.4) of color conservation as

$$i\delta\alpha^a \quad \left(\; A \;\right) = 0 \; . \tag{2.24}$$

In (2.24), the sum of the infinitesimal shifts is obviously the infinitesimal shift of the incoming multi-parton state, and this shift identically vanishes.

A parton system which is fully contracted over parton color indices is SU(N) invariant. Such a system is called a color singlet state. Color conservation is equivalent to the SU(N) invariance of color singlet systems.

Note that if we do not contract with external parton coordinates, the identity (2.24) reads

$$\left(\; A \;\right) = 0 \; , \tag{2.25}$$

with specified external indices $b, c, \ldots, i, j, \ldots$. Let us view the object carrying those indices, $A^{bc\ldots}{}_{i\ldots}{}^{j\ldots}$, as an SU($N$) *tensor*, thus transforming under SU(N) as the product of parton coordinates $g^b g^c \ldots q_i \ldots q^j \ldots$. Equation (2.25) gives an alternative formulation of color conservation:

> All SU(N) tensors (built from the basic legos) are SU(N) invariant tensors.

Exercise 2.4 Check explicitly that the tensors

$$
\overset{i}{\underset{j}{\rule{0pt}{0pt}}}\!\!\!\!\!\!\!\!\!\!\!\!\! \supset \;=\; \delta^j_i \;\;; \qquad \overset{b}{\underset{c}{\rule{0pt}{0pt}}}\!\!\!\!\!\!\!\!\!\! \,=\; \delta_{bc} \;\;, \tag{2.26}
$$

are invariant under color rotations. (One may consider a finite or an infinitesimal rotation, since the invariance under infinitesimal SU(N) transformations obviously implies the invariance under any finite transformation.)

Exercise 2.5 Express the SU(N) invariance of the tensor $(T^a)^j_i$ under *finite* color rotations, to obtain the relation

$$
\tilde{U}_{bc} = 2\,\mathrm{Tr}\,(T^b U T^c U^\dagger)\,, \tag{2.27}
$$

which determines the matrix elements $\tilde{U}_{bc} = \tilde{U}(\alpha)_{bc}$ of a color rotation of angles α in the adjoint representation in terms of its fundamental representation $U = U(\alpha)$.

2.3 SU(N) Irreducible Representations

Using the pictorial expression of color conservation and infinitesimal color rotations, we can approach SU(N) irreducible representations in a fairly intuitive way.

Consider a multi-parton system spanning a color vector space $\mathcal{E}$ of dimension n, and suppose we have at disposal m projectors $\mathbb{P}_i$ constructed from the basic legos and satisfying the conditions $\mathbb{P}_i\,\mathbb{P}_j = 0$ for $i \neq j$ and $\sum_{i=1}^m \mathrm{rank}(\mathbb{P}_i) = n$, implying the completeness relation $\sum_{i=1}^m \mathbb{P}_i = \mathbb{1}_{\mathcal{E}}$. (An explicit case was given in Chap. 1 when proving the Fierz identity, see Exercise 1.2.) We also suppose the projectors to be Hermitian, $\mathbb{P}_i^\dagger = \mathbb{P}_i$.

Let us consider the infinitesimal shift of the parton system under an infinitesimal color rotation (dropping the factor $i\delta\alpha^a$ which is irrelevant to the discussion), and then insert on the left and right the completeness relation:

$$ \tag{2.28} $$

where a dashed vertical line indicates to which subspace of $\mathcal{E}$ (here img($\mathbb{P}_i$) or img($\mathbb{P}_j$)) the corresponding intermediate parton system belongs.

Using color conservation and $\mathbb{P}_i \, \mathbb{P}_j = 0$ for $i \neq j$, only the terms with $i = j$ remain in the double sum. As a consequence, the image space img($\mathbb{P}_i$) of the projector $\mathbb{P}_i$ is invariant under any infinitesimal color rotation, and thus under SU(N). In a basis of $\mathcal{E}$ obtained by joining bases of the invariant subspaces img($\mathbb{P}_i$) (which due to the hermiticity of $\mathbb{P}_i$ are orthogonal to each other), any SU(N) color rotation will be block-diagonal,

$$U_{(\mathcal{E})} = \begin{pmatrix} \square & & & \\ & \square & & \\ & & \ddots & \\ & & & \square \\ & & & \end{pmatrix} . \tag{2.29}$$

If each block cannot be further block-diagonalized, i.e., if the chosen set of projectors is of maximal cardinality, we say that each invariant subspace img($\mathbb{P}_i$) transforms under an irreducible representation (irrep) R_i of SU(N) (the irrep R_i being defined by the set of matrices forming the block i of (2.29)), and that the product $q^i q_j g^a \ldots$ describing the multi-parton system $\{q\bar{q}g \ldots\}$ decomposes into a sum of irreps as:

$$\mathbf{3} \otimes \bar{\mathbf{3}} \otimes \mathbf{8} \otimes \ldots = \overset{m}{\underset{i=1}{\bigoplus}} R_i . \tag{2.30}$$

> To determine all irreps of a parton system, we need to find a maximal, complete set of Hermitian and mutually orthogonal projectors (constructed from the basic legos).

Let's give the pictorial form of the SU(N) generators $T^a(R)$ ($a = 1 \ldots N^2 - 1$) of the representation R associated to the projector $\mathbb{P}_R$ (i.e., acting in the invariant subspace img($\mathbb{P}_R$)):

$$T^a(R) = \begin{array}{c}\includegraphics\end{array} . \tag{2.31}$$

Indeed, $T^a(R)$ defined in this way is a map of img($\mathbb{P}_R$) → img($\mathbb{P}_R$), and $i\delta\alpha^a \, T^a(R)$ acting on a parton state in img($\mathbb{P}_R$) is the infinitesimal shift of this state under the infinitesimal color rotation of angles $\delta\alpha^a$.

Exercise 2.6 Check pictorially that the $T^a(R)$'s satisfy the SU(N) Lie algebra $\left[T^a(R), T^b(R)\right] = i f_{abc} \, T^c(R)$.

From Exercise 2.3 we directly infer that the irrep R is defined by the set of matrices $U_R(\alpha) = e^{i\alpha^a T^a(R)}$.

If the $d_R \times d_R$ matrices $T^a(R)$ ($a = 1 \ldots N^2 - 1$) satisfy the SU(N) Lie algebra and are irreducible (i.e., they cannot all be put in the same block-diagonal form by a change of basis), the matrices $U_R(\alpha) = e^{i\alpha^a T^a(R)}$ define an SU(N) irrep of dimension d_R, acting on objects with d_R components while preserving the group structure.

Note that for $N > 2$, SU(N) irreps do not exist for any dimension d_R. For $N = 3$, the possible dimensions are $d_R = 1, 3, 6, 8, 10, 15 \ldots$, as we will see in the following chapters.

Finally, let us mention that instead of defining an irrep R by the set of matrices $U_R(\alpha)$ acting on img($\mathbb{P}_R$), one may characterize an irrep by its associated *multiplet*. In this manual we define the multiplet μ_R associated to the irrep R as the basis of img($\mathbb{P}_R$) obtained by multiplying the projector $\mathbb{P}_R$ to the left by parton coordinates:

$$[\mu_R]^{i\ldots a}_{j\ldots} \equiv \quad \mathbb{P}_R \quad , \tag{2.32}$$

where the pictorial notation of coordinates was defined in (2.23).

Chapter 3
Diquark Color States, Schur's Lemma and Casimir Charges

Abstract We present a systematic method, called the "tensor method", to find the set of Hermitian projectors on the irreps of a parton system. Here it is explained in the simplest case of a quark pair, but it can be applied to any parton system (as we shall see in following chapters). We also state the extremely useful Schur's lemma, define transition operators between equivalent irreps, and write the pictorial expression of the quadratic Casimir operator of a parton system in a given irrep.

3.1 Irreps of a Quark Pair

Pictorially, a quark pair can be represented as

$$\substack{\circ \!-\!\!\!\to \; i \\ \circ \!-\!\!\!\to \; j} \;\; \equiv \; q^i q^j \, . \tag{3.1}$$

As we saw in Chap. 2, finding a basis of the vector space $\{q^i q^j\} \equiv \mathcal{V} \otimes \mathcal{V}$ (of dimension N^2) where all color rotations (represented by $N^2 \times N^2$ matrices) are block-diagonal (and cannot be further block-diagonalized) amounts to finding a maximal and complete set of Hermitian, mutually orthogonal projectors $\mathbb{P}_i$.

The case of diquarks being very simple, let us immediately give the result for the set of projectors. It is composed of two projectors corresponding to the symmetrizer and anti-symmetrizer (over the two quark indices), given respectively by:

$$P_S = \frac{1}{2}\left(\substack{\to \\ \to} + \times\right) \equiv \substack{\to \\ \to}\boxed{}\substack{\to \\ \to} \; ; \quad P_A = \frac{1}{2}\left(\substack{\to \\ \to} - \times\right) \equiv \substack{\to \\ \to}\blacksquare\substack{\to \\ \to} \, . \tag{3.2}$$

For a system $\{q^i q^j \ldots q^p\}$ made up only of quarks, representation theory tells us that the bases of the SU(N) invariant and irreducible subspaces are given by linear combinations of $q^i q^j \ldots q^p$ having different symmetry properties in the permutation of indices (Hamermesh 1962). In the present case of two quarks, we can build either a totally symmetric or a totally antisymmetric linear combination of $q^i q^j$, leading to the set (3.2) of projectors.

Exercise 3.1 Verify that P_S and P_A form a complete set of Hermitian projectors, which are mutually orthogonal. Show pictorially that their ranks are $N(N+1)/2$ and $N(N-1)/2$, respectively.

Exercise 3.2 The multiplets associated to P_S and P_A (corresponding to bases of img(P_S) and img(P_A), respectively) are given by (see (2.32)):

$$\mu_S^{ij} = \begin{array}{c}\text{⊶⬚⊷}\end{array}\!\!{}^{i}_{j} = \frac{1}{2}\left(q^i q^j + q^j q^i\right), \tag{3.3}$$

$$\mu_A^{ij} = \begin{array}{c}\text{⊶▮⊷}\end{array}\!\!{}^{i}_{j} = \frac{1}{2}\left(q^i q^j - q^j q^i\right). \tag{3.4}$$

Although we know from Chap. 2 that img(P_S) and img(P_A) are invariant under SU(N), verify this explicitly by writing how the multiplets transform under finite color rotations (showing in passing that μ_S^{ij} and μ_A^{ij} are SU(N) tensors of rank 2).

In summary, we have the completeness relation

$$\begin{array}{c}\text{⇉}\end{array} = P_S + P_A = \begin{array}{c}\text{→⬚⊷}\end{array} + \begin{array}{c}\text{→▮⊷}\end{array}, \tag{3.5}$$

and a quark pair, which transforms under SU(N) as the product of two fundamental representations, decomposes into a sum of irreps as

$$\mathbf{3} \otimes \mathbf{3} = \mathbf{6} \oplus \bar{\mathbf{3}}. \tag{3.6}$$

The irreps associated to the projectors P_S and P_A are named after the value of their dimension when $N = 3$ (see Chap. 2), so here **6** and $\bar{\mathbf{3}}$. The irrep of dimension $N(N-1)/2$ associated to P_A is named $\bar{\mathbf{3}}$ and not **3**, because for $N = 3$, μ_A^{ij} does not transform under SU(3) as a quark, but as an antiquark (see the exercise below). This shows that, in general, knowing the dimension of an irrep may not be sufficient to determine it completely. In case of ambiguity, note that an SU(N) irrep is uniquely defined by its associated Hermitian projector.

Exercise 3.3 For $N = 3$, the generators of the irrep of dimension 3 acting on the multiplet μ_A^{ij} are equivalent to $-(T^a)^* = -({}^t T^a)$, i.e., μ_A^{ij} transforms under SU(3) as an antiquark. Prove this by trading the three independent components of μ_A^{ij} for the 3-vector $B_k \equiv \frac{1}{2}\epsilon_{ijk}\mu_A^{ij}$ (with ϵ_{ijk} the Levi-Civita tensor of rank 3) and by evaluating the shift δB_k under an infinitesimal color rotation.

Let us now suppose that we do not know anything about representation theory, and that we therefore do not know from the start the set of Hermitian projectors. We describe below a systematic method to find them (Cvitanović 2008), which we will call the "tensor method".[1] In the qq case, this method is not the most economical, but one advantage of the tensor method is that it can be applied to any parton system composed of quarks, antiquarks and gluons, as we'll see later in this manual.

The "tensor method" (Cvitanović 2008) for finding Hermitian projectors on the irreps of a multi-parton system consists of three steps:

1. Find a basis of operators (built from the basic legos, thus being $SU(N)$ invariant tensors) mapping the multi-parton vector space to itself.
2. Obtain the multiplication table between these operators, and infer the characteristic equation and eigenvalues of the most interesting one(s).
3. Express the projectors on the corresponding eigenspaces in terms of the $SU(N)$ invariant tensors.

Let us apply this general procedure to a quark pair.

1. Basis of tensors

Any operator mapping the vector space $\mathcal{V} \otimes \mathcal{V}$ to itself can be expressed in terms of graphs of the generic form

$$(3.7)$$

Such graphs can be replaced by (linear combinations of) simpler graphs using the following algorithm (Cvitanović 2008; Keppeler 2017), referred to as the "standard algorithm" in the remainder of this manual.

First, we can get rid of any three-gluon vertex appearing in the graph by using the identity (prove it!)

$$(3.8)$$

The graphs then reduce to (linear combinations of) graphs where any internal gluon connects at both ends to quark lines.

Second, every internal gluon line can be removed with the help of the Fierz identity (1.17). So we end up with graphs with four external quark lines (together with airborne quark loops that simply contribute to an irrelevant global factor N^ℓ) and without gluons. There are only two ways to connect the four external quark lines, and this proves that there are only two linearly independent tensors mapping $\mathcal{V} \otimes \mathcal{V}$ to itself, namely,

$$\mathbb{1} \equiv \quad ; \quad X \equiv \quad . \qquad (3.9)$$

---

[1] This is our own terminology. In Cvitanović (2008), this method is referred to as the characteristic equation method.

2. Multiplication table and characteristic equations

The multiplication table of the set $\{\mathbb{1}, X\}$ has only one non-trivial entry,

$$X^2 = \overset{\rightarrow}{\times}\overset{\rightarrow}{\times} = \overset{\longrightarrow}{\longrightarrow} = \mathbb{1} . \tag{3.10}$$

The characteristic equation of the operator X is $X^2 - \mathbb{1} = 0$. The minimal polynomial of X is thus $t^2 - 1 = (t - 1)(t + 1)$, which is split with simple roots. From elementary linear algebra, it follows that X can be diagonalized (which is not a surprise since X is clearly Hermitian) and has eigenvalues $\{x_1, x_2\} = \{1, -1\}$. In some basis of $\{q^i q^j\} \equiv \mathcal{V} \otimes \mathcal{V}$, the matrix representation of X thus reads

$$X = \begin{pmatrix} x_1 & & & \\ & x_1 & & \\ & & x_2 & \\ & & & x_2 \end{pmatrix}. \tag{3.11}$$

3. Projectors on eigenspaces

In the above-mentioned basis, the projectors P_{x_1} and P_{x_2} on the eigenspaces of X are

$$P_{x_1} = \begin{pmatrix} 1 & & & \\ & 1 & & \\ & & 0 & \\ & & & 0 \end{pmatrix} ; \quad P_{x_2} = \begin{pmatrix} 0 & & & \\ & 0 & & \\ & & 1 & \\ & & & 1 \end{pmatrix}. \tag{3.12}$$

Their explicitly $SU(N)$ invariant form follows from the identities $X = x_1 P_{x_1} + x_2 P_{x_2}$ and $\mathbb{1} = P_{x_1} + P_{x_2}$, or more directly from a simple observation of the matrix X given in (3.11):

$$P_{x_1} = \frac{X - x_2 \mathbb{1}}{x_1 - x_2} = \frac{1}{2}(\mathbb{1} + X) = P_S ; \quad P_{x_2} = \frac{X - x_1 \mathbb{1}}{x_2 - x_1} = \frac{1}{2}(\mathbb{1} - X) = P_A . \tag{3.13}$$

We thus recover the projectors (3.2) without any prior knowledge of representation theory. Note that in the above derivation, the resulting projectors satisfy all requirements *by construction*: they are Hermitian and mutually orthogonal, they form a complete set ($P_{x_1} + P_{x_2} = \mathbb{1}$), and they cannot be reduced into a sum of more $SU(N)$ invariant projectors (since this would imply that there are more than two independent tensors mapping $\mathcal{V} \otimes \mathcal{V}$ to itself). Therefore, P_S and P_A must project onto the irreps of a quark pair.

Let's conclude this section with two important remarks:

- Obviously, the number of projectors (i.e. of irreps) cannot exceed the number of independent tensors found in step 1 of the tensor method, $n_{\text{irreps}} \leq n_{\text{tensors}}$. For a quark pair, we have $n_{\text{irreps}} = n_{\text{tensors}}$, but for more complicated systems we may have $n_{\text{irreps}} < n_{\text{tensors}}$. This happens when some of the independent tensors are not Hermitian and therefore cannot contribute to the construction of Hermitian projectors. This will be the case for the qqq system considered in Chap. 4.

- When $n_{\text{irreps}} < n_{\text{tensors}}$, one might naively think that n_{irreps} coincides with the number of independent tensors that are Hermitian, but this is not the case. Indeed, the projectors are linear combinations of some subset of the independent tensors, and the tensors of this subset are thus linear combinations of the projectors. Since the projectors are not only Hermitian but also commute between them, the same must be true for the independent tensors of the subset. We infer that in general, it is the largest subset of *commuting* Hermitian operators found among the independent tensors that is used to construct the projectors, and thus determines the number of irreps.

3.2 Schur's Lemma and Transition Operators

Consider two irreps R_1 and R_2 acting in the respective multi-parton vector spaces $\mathcal{E}_1$ and $\mathcal{E}_2$, and an SU(N) invariant tensor A mapping $\mathcal{E}_1 \rightarrow \mathcal{E}_2$ of the form:

$$A = \;\; \boxed{R_1} \;\; \bigcirc \;\; \boxed{R_2} \;\; , \tag{3.14}$$

where a blob R denotes the Hermitian projector $\mathbb{P}_R$ associated to the irrep R. Since A is an invariant tensor we can use color conservation and write:

$$i\delta\alpha^a \;\; \boxed{R_1} \;\; \bigcirc \;\; \boxed{R_2} \;\; = i\delta\alpha^a \;\; \boxed{R_1} \;\; \bigcirc \;\; \boxed{R_2} \;\; . \tag{3.15}$$

In the l.h.s. of (3.15), the infinitesimal color rotation (strictly speaking, infinitesimal shift) acts in the irrep R_1 (as is pictorially obvious, it maps img($\mathbb{P}_{R_1}$) to itself, see Chap. 2). In the r.h.s., the same color rotation acts in the irrep R_2. For finite color rotations, (3.15) thus reads

$$\forall\, U(\alpha) \in \text{SU}(N)\,, \quad A\, U_{R_1}(\alpha) = U_{R_2}(\alpha)\, A\,. \tag{3.16}$$

The condition (3.16) is the starting assumption for stating Schur's lemma, which consists of two parts (see e.g. Georgi 2000 for a proof), both of which are very important results, useful for simplifying calculations and also for intuition:

- If R_1 and R_2 are inequivalent irreducible representations, then $A = 0$.

This can be proven by showing that if $A \neq 0$, A must be an invertible square matrix, so for any given rotation angles α we have $U_{R_1}(\alpha) = A^{-1}U_{R_2}(\alpha)A$. This means that R_1 and R_2 are simply related by a change of basis, which is the definition of *equivalent* representations. Viewing the tensor A represented in (3.14) as the "evolution" of a parton system, we see that in absence of interaction with external color fields, a parton system may change its composition, but always remains in equivalent irreps.

- if $\mathcal{E}_1 = \mathcal{E}_2$ (i.e., the initial and final parton systems in (3.14) are the same) and $R_1 = R_2 \equiv R$, then A is proportional to the identity operator in the irrep R given by $\mathbb{1}_R = \mathbb{P}_R$:

$$\text{(diagram)} \; = \; c \; \text{(diagram)} \; = \; c\,\mathbb{P}_R . \qquad (3.17)$$

The latter equation can be interpreted as follows (Georgi 2000). Suppose we prepare an incoming multiplet μ_R in irrep R and try to mix the basis states of this multiplet with the help of an invertible matrix (the white blob in the l.h.s. of (3.17)) which is thus a map of $\text{img}(\mathbb{P}_R) \to \text{img}(\mathbb{P}_R)$. Due to Schur's lemma (3.17), up to an overall factor we get exactly the same states. The basis states of a multiplet are uniquely defined.

Let us mention that when $R_1 \neq R_2$, a non-zero tensor A of the form (3.14) is called a *transition operator* (for the transition $R_1 \to R_2$). In this case, the first part of Schur's lemma can be reformulated as:

There is a transition operator A between R_1 and $R_2 \neq R_1$ *if and only if* R_1 and R_2 are equivalent irreps, and A is then a similarity transformation between the two irreps, $U_{R_1}(\alpha) = A^{-1}U_{R_2}(\alpha)A$.

Exercise 3.4 When $R_1 \neq R_2$ are equivalent irreps, show that the transition operator between R_1 and R_2 is uniquely defined (up to an overall factor).

3.3 Casimir Charges

In Chap. 2 we gave the pictorial expression of $SU(N)$ generators $T^a(R)$ in the irrep R, see Eq. (2.31). The (quadratic) Casimir operator in the irrep R is defined by $T^a(R)T^a(R)$, which from Schur's lemma (3.17) must be proportional to $\mathbb{P}_R$, with a proportionality coefficient named the Casimir charge C_R:

$$T^a(R)T^a(R) \; = \; \text{(diagram)} \; = \; C_R \; \text{(diagram)} . \qquad (3.18)$$

Note that unlike the generators $T^a(R)$, the Casimir operator commutes with all SU(N) transformations. In Chap. 1, we already met the Casimir operators $T^a T^a = C_{_F} \mathbb{1}_V$ and $t^a t^a = C_{_A} \mathbb{1}_{\mathcal{A}}$ in the fundamental (quark) and adjoint (gluon) representations.

Exercise 3.5 Show that the global Casimir charge C_R of a color state R of two partons (of individual Casimir charges C_1 and C_2) can be written as

$$C_R = C_1 + C_2 + V_{12}(R) , \tag{3.19}$$

with the color "interaction potential" $V_{12}(R)$ of the parton pair in irrep R defined by (the dashed lines representing partons 1 and 2):

$$V_{12}(R)\,\mathbb{P}_R \; \equiv \; -2 \quad \mathbb{P}_R \; = \; -2 \quad \left(\mathbb{P}_R\right) . \tag{3.20}$$

What is the generalization of (3.19) to a system of $n > 2$ partons?

Exercise 3.6 Calculate the Casimir charges of the two diquark irreps **6** and $\bar{\mathbf{3}}$ (associated with the projectors P_S and P_A given in (3.2)) as a function of N. In which color state is the color interaction potential attractive?

Chapter 4
Color States of $q\bar{q}, qg$ and qqq Systems

Abstract We systematically apply the tensor method described in Chap. 3 for a pair
of quarks to find the irreps of a few other simple systems: $q\bar{q}$, qg, and the qqq
system. The latter is the simplest system for which the number of irreps is smaller
than the number of independent tensors mapping the parton system color space, thus
providing a simple illustration of transition operators between equivalent irreps.

4.1 Quark-Antiquark Pair

As a warming up, let us consider $q\bar{q}$ pairs. The tensor method for finding how $\mathbf{3} \otimes \bar{\mathbf{3}}$
decomposes into a sum of irreps is as follows.

1. Basis of tensors

Similarly to what was done for qq pairs in Chap. 3, using the standard algorithm (see
(3.7) and the discussion below) one can obtain a basis of tensors mapping $\mathcal{V} \otimes \overline{\mathcal{V}}$
to itself. For instance, one can choose:

$$\mathbb{1} \equiv \overset{\longleftarrow}{\underset{\longrightarrow}{}} \ ; \quad S \equiv \big\}\big\{ . \tag{4.1}$$

2. Multiplication table and characteristic equations

The multiplication table of the set (4.1) has only one non-trivial entry:

$$S^2 = \big\}\big[\big]\big\{ = N\big\}\big\{ = NS . \tag{4.2}$$

Thus, S has for minimal polynomial $t^2 - Nt$, and for eigenvalues $\{s_1, s_2\} = \{N, 0\}$.
In some basis of $\{q^i \bar{q}_j\} \equiv \mathcal{V} \otimes \overline{\mathcal{V}}$, we have $S = \text{diag}(s_1 \ldots s_1, s_2 \ldots s_2)$.

3. Projectors on eigenspaces

The projectors on the eigenspaces of S read

$$P_{s_1} = \frac{S - s_2\mathbb{1}}{s_1 - s_2} = \frac{S}{N} = \frac{1}{N}\big]\big\{ \equiv \mathbb{P}_1 , \tag{4.3}$$

© The Author(s), under exclusive license to Springer Nature Switzerland AG 2024
S. Peigné, *Color in QCD*, SpringerBriefs in Physics,
https://doi.org/10.1007/978-3-031-53681-6_4

$$P_{s_2} = \frac{S - s_1 \mathbb{1}}{s_2 - s_1} = \mathbb{1} - \frac{S}{N} = \overrightarrow{\quad\longleftarrow\quad} - \frac{1}{N} \left. \rule{0pt}{14pt}\right] \left[\rule{0pt}{14pt}\right. = 2 \, \text{\ding{mmm}} \equiv \mathbb{P}_8 \; . \qquad (4.4)$$

We thus recover the projectors of ranks 1 and $N^2 - 1$ encountered when proving the Fierz identity (1.17) in Chap. 1 (see Exercise 1.2). Recall that the tensor method guarantees that these projectors correspond to *irreducible* SU(N) representations, which are thus named **1** and **8**. (The invariant subspace img($\mathbb{P}_8$) must be irreducible, otherwise one would have found more than two independent tensors in step 1 of the procedure).

The irrep **1** is called the trivial (or singlet) representation of SU(N), and the irrep **8** is equivalent to the adjoint (gluon) representation, as is obvious from the pictorial form of the projector $\mathbb{P}_8$ and Schur's lemma. In summary, a $q\bar{q}$ pair decomposes into a sum of irreps as

$$\mathbf{3} \otimes \bar{\mathbf{3}} = \mathbf{8} \oplus \mathbf{1} \, , \qquad (4.5)$$

the Fierz identity (1.17) being nothing but the completeness relation $\mathbb{1} = \mathbb{P}_8 + \mathbb{P}_1$.

Exercise 4.1 Write the multiplets (generally defined by (2.32)) of $q^i q_j$ associated to $\mathbb{P}_1$ and $\mathbb{P}_8$, and check explicitly that they transform as expected under SU(N).

4.2 Quark-Gluon Pair

4.2.1 Derivation of qg Irreps

Finding the color states of a quark-gluon pair using the tensor method is fairly straightforward (Cvitanović 2008). Since you get used to it, several steps of the derivation are left as exercises.

1. Basis of tensors

For a basis of tensors mapping $\mathcal{V} \otimes \mathcal{A}$ to itself (recall that $\mathcal{A} \equiv \{g^a\}$ is the gluon vector space), one can take:

$$I \equiv \overrightarrow{\text{\ding{mmm}}} \; ; \quad A \equiv \text{\ding{mmm}} \; ; \quad B \equiv \text{\ding{mmm}} \; . \qquad (4.6)$$

Exercise 4.2 Prove it by using the standard algorithm (see Chap. 3), and paying attention to quark loops.

2. Multiplication table and characteristic equations

The non-trivial entries of the multiplication table are:

$$A^2 = C_F A \; ; \quad AB = BA = -\frac{1}{2N} A \; ; \quad B^2 = \frac{1}{4} I - \frac{1}{2N} A \, . \qquad (4.7)$$

Exercise 4.3 Derive the relations (4.7) in a pictorial way. How many irreps can we expect from the set of tensors (4.6), and why? Infer the minimal polynomial of B, and explain why there is a basis of $\{q^i g^a\} \equiv \mathcal{V} \otimes \mathcal{A}$ where B is represented by the matrix $B = \text{diag}(b_1 \ldots b_1, b_2 \ldots b_2, b_3 \ldots b_3)$, with $\{b_1, b_2, b_3\} = \{\frac{1}{2}, -\frac{1}{2}, -\frac{1}{2N}\}$.

3. Projectors on eigenspaces

The projectors on the eigenspaces of B can be expressed in the SU(N) invariant form:

$$P_{b_1} = \frac{1}{2} \overset{\longrightarrow}{\text{\tiny mmm}} - \frac{1}{N+1} \overset{}{\text{\tiny}} + \overset{}{\text{\tiny}} \; , \qquad (4.8)$$

$$P_{b_2} = \frac{1}{2} \overset{\longrightarrow}{\text{\tiny mmm}} - \frac{1}{N-1} \overset{}{\text{\tiny}} - \overset{}{\text{\tiny}} \; , \qquad (4.9)$$

$$P_{b_3} = \frac{1}{C_{\text{F}}} \overset{}{\text{\tiny}} \; . \qquad (4.10)$$

Exercise 4.4 Obtain the projectors (4.8)–(4.10) using the formula

$$P_{b_i} = \frac{(B - b_j)(B - b_k)}{(b_i - b_j)(b_i - b_k)} \; , \qquad (4.11)$$

where P_{b_i} is the projector on the eigenspace associated to the eigenvalue b_i, and b_j, b_k are the two other eigenvalues. (The formula (4.11) follows from a mere observation of B in its diagonal form.) Note that the projectors P_{b_i} are sometimes called the "Frobenius covariants" of B.

By construction, the projectors (4.8)–(4.10) form a complete (and maximal) set of Hermitian, mutually orthogonal projectors, which thus project on the irreps of a quark-gluon pair.

Exercise 4.5 Denoting by α the qg irreps associated to the projectors ordered as in (4.8)–(4.10), show that the irrep dimensions K_α, Casimir charges C_α, and color interaction potentials $V_\alpha \equiv V_{qg}(\alpha)$ (see (3.20)) are given by

$$K_\alpha = \left\{ \frac{1}{2}N(N+2)(N-1), \; \frac{1}{2}N(N-2)(N+1), \; N \right\} , \qquad (4.12)$$

$$C_\alpha = \left\{ \frac{(N+1)(3N-1)}{2N}, \; \frac{(N-1)(3N+1)}{2N}, \; C_{\text{F}} \right\} , \qquad (4.13)$$

$$V_\alpha = \{1, \; -1, \; -N\} . \qquad (4.14)$$

Using (4.12) and naming as usual the SU(N) irreps by their dimensions for $N = 3$, the color decomposition of a qg pair can be expressed as

$$\mathbf{8} \otimes \mathbf{3} = \mathbf{15} \oplus \bar{\mathbf{6}} \oplus \mathbf{3} \,, \tag{4.15}$$

or equivalently by the completeness relation

$$\text{———} = \mathbb{P}_{15} + \mathbb{P}_{\bar{6}} + \mathbb{P}_3 \,, \tag{4.16}$$

where the qg projectors $\mathbb{P}_\alpha$ for $\alpha = \mathbf{15}, \bar{\mathbf{6}}$ and $\mathbf{3}$ are given by (4.8), (4.9) and (4.10), respectively.

Exercise 4.6 Verify that the projectors $\mathbb{P}_{15}$ and $\mathbb{P}_{\bar{6}}$ can be rewritten as

$$\mathbb{P}_{15} = 2 \left[\, \text{———} - \mathbb{P}_3 \, \right] \text{———} \,, \tag{4.17}$$

$$\mathbb{P}_{\bar{6}} = 2 \left[\, \text{———} - \mathbb{P}_3 \, \right] \text{———} \,, \tag{4.18}$$

where the diquark symmetrizer and antisymmetrizer were defined in (3.2).

The irrep associated to the projector (4.9) has been named $\bar{\mathbf{6}}$ (and not simply $\mathbf{6}$) without any explanation. To free yourself from the anxiety that has probably arisen, do the following exercise:

Exercise 4.7 Show that for $N = 3$, the irrep associated to (4.9) is equivalent to the *complex conjugate* of the irrep $\mathbf{6}$ appearing in the decomposition $\mathbf{3} \otimes \mathbf{3} = \mathbf{6} \oplus \bar{\mathbf{3}}$ of a quark pair (see (3.6)). *Hint: use (4.17)–(4.18) and remember Schur's lemma.*

To avoid any blunders, note that when $N > 3$, the SU(N) qq irrep $\mathbf{6}$ and qg irrep $\bar{\mathbf{6}}$ have nothing in common, as can be seen by comparing their dimensions and Casimir charges as a function of N.

4.2.2   Why the Tensor Method Cannot Fail

In the above derivation of the qg irreps using the tensor method, we first identified a basis of three tensors mapping $\mathcal{V} \otimes \mathcal{A} \to \mathcal{V} \otimes \mathcal{A}$. Thus, we know from the start that the tensors $\{I, A, A^2, A^3\}$ are linearly related, as well as the tensors $\{I, B, B^2, B^3\}$. This implies that both A and B have a minimal polynomial of degree $d \leq 3$. For the operator B, we have exactly $d = 3$ (see Exercise 4.3), and since B is diagonalizable, B must have three *distinct* eigenvalues. The eigenspaces of B thus separate the space $\mathcal{V} \otimes \mathcal{A}$ into the three irreducible SU(N) invariant subspaces we are looking for.

For more complicated parton systems, among the n independent commuting Hermitian tensors used to construct n irreps, there may be no tensor with a minimal polynomial of degree $d = n$ and therefore n distinct eigenvalues. (Or, if there is such a tensor, finding exactly which one and deriving its characteristic equation could be very tedious.) In those cases, the common procedure is to first use a tensor with a simple characteristic equation, with a minimal polynomial of degree $d < n$, thus having d eigenspaces separating the complete space into d invariant subspaces. Then, choosing one of the remaining tensors and studying its action in each of those subspaces will further separate the space. This must be repeated until n invariant subspaces (or equivalently n projectors) are found.

To illustrate this procedure, let us construct the qg irreps by first using the operator A, of characteristic equation $A^2 = C_F A$ and eigenvalues $\{a_1, a_2\} = \{0, C_F\}$. In some basis of $\mathcal{V} \otimes \mathcal{A}$, we have $A = \operatorname{diag}(a_1 \ldots a_1, a_2 \ldots a_2)$, and the projectors on the eigenspaces of A thus read:

$$P_{a_1} = \frac{A - a_2}{a_1 - a_2} = I - \frac{A}{C_F} \quad ; \quad P_{a_2} = \frac{A - a_1}{a_2 - a_1} = \frac{A}{C_F} . \tag{4.19}$$

By construction, $\{P_{a_1}, P_{a_2}\}$ is a complete set of Hermitian, mutually orthogonal projectors (but not yet maximal). As a consequence, the eigenspaces $\operatorname{img}(P_{a_1})$ and $\operatorname{img}(P_{a_2})$ are SU(N) invariant subspaces, and $\mathcal{V} \otimes \mathcal{A} = \operatorname{img}(P_{a_1}) \oplus \operatorname{img}(P_{a_2})$. Using $I = P_{a_1} + P_{a_2}$ and $A = a_1 P_{a_1} + a_2 P_{a_2}$, we see that this first separation of the complete space achieved by the tensor A amounts to trading the tensors I, A for the projectors P_{a_1}, P_{a_2}. To proceed with the construction of the missing projector, we can thus use the set of tensors $\{P_{a_1}, P_{a_2}, B\}$.

Since A and B commute, $BP_{a_i} = P_{a_i}B$, i.e., B leaves each subspace $\operatorname{img}(P_{a_i})$ invariant, and we can consider the restriction BP_{a_i} of B to $\operatorname{img}(P_{a_i})$. In $\operatorname{img}(P_{a_1})$, the tensors P_{a_1}, BP_{a_1}, $B^2 P_{a_1}$ must be linearly related (otherwise, with P_{a_2}, we would have four independent tensors). This means that BP_{a_1} has a minimal polynomial of degree $d_1 \leq 2$ and thus $d_1 \leq 2$ eigenvalues and corresponding eigenspaces. Similarly, in $\operatorname{img}(P_{a_2})$ the tensors P_{a_2}, BP_{a_2}, $B^2 P_{a_2}$ are linearly related and BP_{a_2} has $d_2 \leq 2$ eigenspaces. Since each eigenspace provides an SU(N) invariant subspace, and in the present case we must find $n = 3$ irreps, we have a fortiori either $d_1 = 2$ and $d_2 = 1$, or vice versa. Using (4.7) we readily verify that $B^2 P_{a_1} = \frac{1}{4} P_{a_1}$ and $BP_{a_2} = -\frac{1}{2N} P_{a_2}$, i.e., $d_1 = 2$ and $d_2 = 1$.

The tensor BP_{a_2} is proportional to the identity of the space $\operatorname{img}(P_{a_2})$, so B does not split further this subspace, and the projector P_{a_2} thus corresponds to a qg irrep. (Indeed, it coincides with the projector (4.10) on the eigenspace of B for the eigenvalue $b_3 = -\frac{1}{2N}$ found previously.) As for the restriction BP_{a_1} of B to $\operatorname{img}(P_{a_1})$, it has two eigenvalues $\{b_1, b_2\} = \{\frac{1}{2}, -\frac{1}{2}\}$. In some basis of $\operatorname{img}(P_{a_1})$, $BP_{a_1} = \operatorname{diag}(b_1 \ldots b_1, b_2 \ldots b_2)$, giving directly the projectors on the associated eigenspaces in invariant form:

$$P_{b_1} = \frac{B - b_2}{b_1 - b_2} P_{a_1} = \left(B + \frac{I}{2} \right)\left(I - \frac{A}{C_F} \right), \tag{4.20}$$

$$P_{b_2} = \frac{B - b_1}{b_2 - b_1} P_{a_1} = \left(-B + \frac{I}{2} \right)\left(I - \frac{A}{C_F} \right). \tag{4.21}$$

The tensor B separates $\mathrm{img}(P_{a_1})$ into two parts, $\mathrm{img}(P_{a_1}) = \mathrm{img}(P_{b_1}) \oplus \mathrm{img}(P_{b_2})$. Using $P_{a_1} = P_{b_1} + P_{b_2}$ and $B P_{a_1} = b_1 P_{b_1} + b_2 P_{b_2}$, it is clear that the latter separation amounts to trading the tensors P_{a_1}, $B P_{a_1}$ for P_{b_1}, P_{b_2}. After considering the action of the tensor B, the set $\{P_{a_1}, P_{a_2}, B\}$ can thus be replaced by $\{P_{b_1}, P_{b_2}, P_{a_2}\}$, which thus provides the Hermitian projectors on the three qg irreps.

Exercise 4.8 Verify that the expressions (4.20)–(4.21) coincide with the expressions (4.8)–(4.9) found previously by separating the full space using the sole tensor B.

The above discussion should make it clear why the tensor method *must* lead to the decomposition of the complete space into a sum of irreps, even if we cannot identify a tensor that achieves this decomposition in a single step (like the operator B in the present case, see Sect. 4.2.1).

The procedure of separation of the full space into smaller and smaller invariant subspaces will be systematically applied in the following, for instance to study the qqq system (see Sect. 4.3) or the gluon pair (see Chap. 5). Note that this method naturally yields the projectors in a factorized form (see (4.20)–(4.21)), which is very convenient in some cases, especially for checking the orthogonality of projectors. In practice, it is always a good idea to have the expanded and factorized forms of projectors at hand.

4.3 System or Three Quarks

Here we consider the qqq system (see also Cvitanović 2008; Keppeler 2017), which is important to study once in a lifetime. Indeed, for $N = 3$ the decomposition of qqq into multiplets contains the color singlet *baryons* of the real world. Moreover, it is a simple case allowing to discuss transition operators.

4.3.1 Decomposition of qqq into Irreps

Let us apply the tensor method to a system of three quarks.

1. Basis of tensors

Using the standard algorithm, one finds the following basis of six tensors mapping $\mathcal{V} \otimes \mathcal{V} \otimes \mathcal{V} \equiv \mathcal{V}^{\otimes 3} \to \mathcal{V}^{\otimes 3}$:

Table 4.1 Multiplication table of the set of tensors (4.22). To read this table correctly, note that $\sigma X_2 = X_3$, $X_2 \sigma = X_1 \ldots$, each entry being easily checked pictorially

.	I	σ	τ	X_1	X_2	X_3
I	I	σ	τ	X_1	X_2	X_3
σ	σ	τ	I	X_2	X_3	X_1
τ	τ	I	σ	X_3	X_1	X_2
X_1	X_1	X_3	X_2	I	τ	σ
X_2	X_2	X_1	X_3	σ	I	τ
X_3	X_3	X_2	X_1	τ	σ	I

$$X_1 \equiv \overrightarrow{\times} \; ; \quad X_2 \equiv \times\!\!\!\!\times \quad X_3 \equiv \times\!\!\!\!\underset{\longrightarrow}{} \; , \tag{4.22}$$

corresponding to the elements of the permutation group on three objects S_3.

This is the first example we encounter where some of the independent tensors (here σ and $\tau = \sigma^\dagger$) are not Hermitian. Thus, not all six tensors can be used to construct Hermitian projectors, and for the qqq system we must have $n_{\text{irreps}} < n_{\text{tensors}}$ (see the remarks in the end of Sect. 3.1).

2. Multiplication table

The multiplication table of the tensors (4.22) is identical to that of the group S_3 and shown in Table 4.1.

In order to determine the qqq irreps and associated projectors, we first need to find a maximal set of *commuting* Hermitian tensors using the set (4.22). From the above table we easily verify the following points:

- The tensors $\sigma + \tau$ and $\Sigma \equiv X_1 + X_2 + X_3$ are Hermitian and commute with every tensor of the set (4.22). We can thus take the tensors $I, \sigma + \tau, \Sigma$ in the set we are looking for.
- The tensors X_1, X_2, X_3 are Hermitian, but do not commute with each other. So only one of the X_i's can be added to the set. Let us choose X_3.

We therefore choose $\{I, \sigma + \tau, \Sigma, X_3\}$ as a maximal set from which we can find four irreps and the associated projectors. The tensors $\sigma - \tau$ and $X_2 - X_1$, which complete the latter set to form a basis of six tensors mapping $\mathcal{V}^{\otimes 3} \to \mathcal{V}^{\otimes 3}$, are not necessary for the construction of irreps. Their interpretation will be given in Sect. 4.3.2.

In order to proceed with the explicit construction of the projectors, we observe that the operator $\sigma + \tau$ has a simple characteristic equation:

$$(\sigma + \tau)^2 = \sigma + \tau + 2I \; . \tag{4.23}$$

Its minimal polynomial is thus $t^2 - t - 2 = (t+1)(t-2)$, and in some basis of $\{q^i q^j q^k\} \equiv \mathcal{V}^{\otimes 3}$, it is represented by the matrix

$$
\sigma + \tau = \begin{pmatrix} -1 \\ & \ddots \\ & & -1 \\ & & & 2 \\ & & & & \ddots \\ & & & & & 2 \end{pmatrix} . \tag{4.24}
$$

3. Projectors on eigenspaces

The projectors on the eigenspaces of $\sigma + \tau$ associated to the eigenvalues -1 and 2 are given by

$$
P_{(-1)} = \frac{1}{3}(2I - \sigma - \tau) \;; \quad P_{(2)} = \frac{1}{3}(I + \sigma + \tau) . \tag{4.25}
$$

By construction, these projectors are Hermitian and mutually orthogonal, and satisfy the completeness relation $P_{(-1)} + P_{(2)} = I = \mathbb{1}_{\mathcal{V}^{\otimes 3}}$. Thus, they split the vector space $\mathcal{V}^{\otimes 3}$ into two $SU(N)$ invariant subspaces. Since we need to find four irreps, we continue to separate the space by considering the action of another tensor in our set of four, e.g. X_3, in each subspace already found.

Let us start with the action of X_3 in $\text{img}(P_{(-1)})$. Since $X_3^2 = I$, quite trivially the minimal polynomial of X_3 in any subspace is $t^2 - 1$. In some basis of $\text{img}(P_{(-1)})$ the operator X_3 is thus of the form

$$
X_3|_{\text{img}(P_{(-1)})} = \begin{pmatrix} 1 \\ & \ddots \\ & & 1 \\ & & & -1 \\ & & & & \ddots \\ & & & & & -1 \end{pmatrix} . \tag{4.26}
$$

The projectors on the respective eigenspaces read

$$
P_\pm = \frac{I \pm X_3}{2} \, P_{(-1)} = \frac{1}{6}[2I - \sigma - \tau \pm (2X_3 - X_1 - X_2)] . \tag{4.27}
$$

Applying the same reasoning in $\text{img}(P_{(2)})$, we obtain two other projectors (on the two eigenspaces of X_3 restricted to $\text{img}(P_{(2)})$):

$$
\tilde{P}_\pm = \frac{I \pm X_3}{2} \, P_{(2)} = \frac{1}{6}[I + \sigma + \tau \pm (X_1 + X_2 + X_3)] . \tag{4.28}
$$

We have found four projectors satisfying all requirements, and thus the four irreps of the qqq system.

The projectors $\tilde{P}_\pm$ coincide respectively with the symmetrizer and antisymmetrizer over the three quark indices, defined and denoted pictorially as (Cvitanović 2008, Keppeler 2017):

$$\tilde{P}_+ \equiv \frac{1}{3!} \sum_{\pi \in S_3} \pi \equiv \quad\begin{array}{c}\end{array}\quad , \tag{4.29}$$

$$\tilde{P}_- = \frac{1}{3!} \sum_{\pi \in S_3} \mathrm{sign}(\pi)\, \pi \equiv \quad\begin{array}{c}\end{array}\quad , \tag{4.30}$$

where the sums are over all permutations π of S_3 (i.e., over all tensors of the set (4.22)), and $\mathrm{sign}(\pi)$ denotes the signature of the permutation (with $\mathrm{sign}(\pi) = +1$ for $\pi = I, \sigma, \tau$ and $\mathrm{sign}(\pi) = -1$ for $\pi = X_1, X_2, X_3$).

Exercise 4.9 Show that the projectors $P_\pm$ given by (4.27) can be written as

$$P_+ = \frac{4}{3} \quad\begin{array}{c}\end{array}\quad , \tag{4.31}$$

$$P_- = \frac{4}{3} \quad\begin{array}{c}\end{array}\quad . \tag{4.32}$$

These projectors correspond to mixed symmetries in the three quark indices.

Exercise 4.10 Find the dimensions of the four qqq irreps as a function of N.

Naming as usual the irreps by their dimensions for $N = 3$, the irreps associated with the projectors $\mathbb{P}_\alpha \equiv \{\tilde{P}_-, P_+, P_-, \tilde{P}_+\}$ are labelled by $\alpha = \{\mathbf{1}, \mathbf{8}_+, \mathbf{8}_-, \mathbf{10}\}$. The decomposition of qqq into a sum of SU(N) irreps thus reads

$$\mathbf{3} \otimes \mathbf{3} \otimes \mathbf{3} = \mathbf{1} \oplus \mathbf{8}_+ \oplus \mathbf{8}_- \oplus \mathbf{10} . \tag{4.33}$$

Exercise 4.11 For $N = 3$, the qqq irrep $\mathbf{1}$ is the representation of the color singlet baryons, whose Casimir charge must vanish. Calculate the Casimir charge of the SU(N) irrep $\mathbf{1}$ for general N.

4.3.2 Transition Operators

Among the six independent tensors contributing to the construction of all possible maps of $\mathcal{V}^{\otimes 3} \to \mathcal{V}^{\otimes 3}$, the chosen set $\{I, \sigma + \tau, \Sigma, X_3\}$ (which can be traded for the four $\mathbb{P}_\alpha$'s defined above) allows one to build four irreps. What is the interpretation of the remaining tensors $\sigma - \tau$ and $X_2 - X_1$?

Let us repeat the counting of independent tensors as follows. Using the completeness relation $I = \sum_\alpha \mathbb{P}_\alpha$, any map of $\mathcal{V}^{\otimes 3}$ can be put in the form

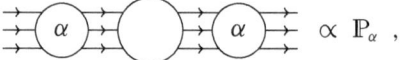

$$\text{(4.34)}$$

From Schur's lemma, in (4.34) only the terms for which α and β are equivalent irreps can contribute. Since equivalent irreps must at least have the same dimension, all possible maps are encompassed by the six structures:

$$\propto \mathbb{P}_\alpha \ ,$$

$$\text{(4.35)}$$

Since the $\mathbb{P}_\alpha$'s are linearly related to $\{I, \sigma + \tau, \Sigma, X_3\}$, we infer that the last two structures of (4.35) span the same set of maps as $\{\sigma - \tau, X_2 - X_1\}$. In particular, they must be non-zero for some choice of the middle blob, thus defining *transition operators* for the transitions $\mathbf{8}_+ \leftrightarrow \mathbf{8}_-$. (If it was not already obvious, this proves that $\mathbf{8}_+$ and $\mathbf{8}_-$ are equivalent irreps). The operators $\sigma - \tau$ and $X_2 - X_1$ are responsible for the transitions $\mathbf{8}_+ \leftrightarrow \mathbf{8}_-$.

We have seen in Chap. 3 that a transition operator is uniquely defined, see Exercise 3.4. In order to find the $\mathbf{8}_+ \leftrightarrow \mathbf{8}_-$ transition operators, we may insert the tensor $\sigma - \tau$ in the middle blob of the last two structures of (4.35), which yields:

$$Q_a \equiv \mathbb{P}_{\mathbf{8}_+}(\sigma - \tau)\, \mathbb{P}_{\mathbf{8}_-} = \quad\quad\quad\quad\quad , \quad\quad \text{(4.36)}$$

$$Q_b \equiv Q_a^\dagger = \mathbb{P}_{\mathbf{8}_-}(\tau - \sigma)\, \mathbb{P}_{\mathbf{8}_+} = \quad\quad\quad\quad\quad , \quad\quad \text{(4.37)}$$

for the $\mathbf{8}_- \to \mathbf{8}_+$ and $\mathbf{8}_+ \to \mathbf{8}_-$ transition operators, respectively. Note that Q_a and Q_b are nilpotent operators, $Q_a^2 = Q_b^2 = 0$, as is the case for any transition operator between irreps whose associated projectors are orthogonal.

To familiarize yourself with transition operators, here are a few exercises:

Exercise 4.12 Express Q_a and Q_b as linear combinations of $\sigma - \tau$ and $X_2 - X_1$.

Exercise 4.13 Check that $Q_a Q_b = 3\mathbb{P}_{\mathbf{8}_+}$ and $Q_b Q_a = 3\mathbb{P}_{\mathbf{8}_-}$, i.e., $Q_a Q_b$ and $Q_b Q_a$ are proportional to the identity operators in the subspaces $\mathrm{img}(\mathbb{P}_{\mathbf{8}_+})$ and $\mathrm{img}(\mathbb{P}_{\mathbf{8}_-})$, respectively.

Exercise 4.14 Find all *unitary* similarity transformations between $\mathbf{8}_-$ and $\mathbf{8}_+$.

Exercise 4.15 If we had chosen X_2 instead of X_3 in the set of four commuting Hermitian tensors (being thus $\{I, \sigma + \tau, \Sigma, X_2\}$), what would be the expressions of the projectors on the four irreps? Denoting the new decomposition as $\mathbf{3} \otimes \mathbf{3} \otimes \mathbf{3} = \mathbf{1} \oplus \mathbf{8}'_+ \oplus \mathbf{8}'_- \oplus \mathbf{10}$, compare the multiplets associated to the irreps $\mathbf{8}'_+$ and $\mathbf{8}'_-$ to those associated to $\mathbf{8}_+$ and $\mathbf{8}_-$. (Recall that a multiplet has been defined by (2.32).)

This last exercise illustrates the following general property:

> When $m \geq 2$ equivalent irreps of Hermitian projectors $\mathbb{P}_i$ ($i = 1 \ldots m$) appear in the decomposition of a multi-parton system, and no specific symmetry is imposed on the $\mathbb{P}_i$'s (such as requiring $\mathbb{P}_i$ to be either symmetric or antisymmetric under permutation of two given identical partons), the projectors $\mathbb{P}_i$ are not uniquely defined separately, only their sum is. In the context of the tensor method, this results from the freedom of choice of the set of commuting Hermitian tensors used to construct the projectors.

Chapter 5
Color States of a Gluon Pair

Abstract We first introduce the "star vertex" and derive new simple rules involving this vertex. Using the tensor method, we then derive the SU(N) irreps of a gluon pair and their associated projectors, starting with a detailed discussion of the basis of tensors used to construct them. Having seven irreps (for $N > 3$) or six irreps (for $N = 3$), the gg system is richer than the systems considered in previous chapters, making the case of a gluon pair ideal for sharpening your birdtrack drawing skills.

5.1 The Star

We start by playing a little with the SU(N) invariant tensor defined by

$$\text{[birdtrack diagram]} = 2 \left(\text{[diagram]} - \text{[diagram]} \right), \tag{5.1}$$

and denoted in index notation by the d_{abc} symbol:

$$a \text{[birdtrack diagram]} c \equiv d_{abc}. \tag{5.2}$$

As we can see, the "star" is built from the basic legos (as any SU(N) invariant tensor) and is therefore not really a novelty, but this effective three-leg vertex is quite convenient when studying systems with several gluons.

Exercise 5.1 Check that the star (5.1) is real and symmetric under the exchange of two lines (i.e., d_{abc} is a real, totally symmetric tensor), and trivially satisfies

$$\text{[diagram]} = 0 \; ; \quad \text{[diagram]} = 0. \tag{5.3}$$

Exercise 5.2 (Watch out for the big conceptual breakthrough: it will now be implicit that the equations mentioned in the exercises must be proven!)

© The Author(s), under exclusive license to Springer Nature Switzerland AG 2024
S. Peigné, *Color in QCD*, SpringerBriefs in Physics,
https://doi.org/10.1007/978-3-031-53681-6_5

$$\text{(diagram)} = \text{(diagram)} + \text{(diagram)} - \frac{1}{N} \text{(diagram)} . \tag{5.4}$$

It follows from (5.4) that the *anti*-commutators of SU(N) generators are fully determined by the d_{abc}'s, namely,

$$\{T^a, T^b\} = d_{abc} T^c + \frac{1}{N} \delta_{ab} . \tag{5.5}$$

Note that this equation is a relation between $N \times N$ matrices, so there is an implicit factor $\mathbb{1}_{N \times N} = \mathbb{1}_V$ in the last term.

Exercise 5.3

$$\text{(diagram)} = \frac{N^2 - 4}{2N} \text{(diagram)} . \tag{5.6}$$

As usual, the SU(N) invariance of the d_{abc} tensor can be expressed pictorially by color conservation (see Chap. 2):

$$\text{(diagram)} = \text{(diagram)} + \text{(diagram)} . \tag{5.7}$$

Exercise 5.4

$$\text{(diagram)} = \text{(diagram)} = \text{(diagram)} = \frac{N}{2} \text{(diagram)} . \tag{5.8}$$

For further use, let us collect Eqs. (3.8) and (5.1) and their reciprocal relations:

$$\text{(diagram)} = 2 \left(\text{(diagram)} + \text{(diagram)} \right) , \tag{5.9}$$

$$\text{(diagram)} = 2 \left(\text{(diagram)} - \text{(diagram)} \right) , \tag{5.10}$$

$$\text{(diagram)} = \frac{1}{4} \left(\text{(diagram)} + \text{(diagram)} \right) , \tag{5.11}$$

$$\text{(diagram)} = \frac{1}{4} \left(\text{(diagram)} - \text{(diagram)} \right) . \tag{5.12}$$

Exercise 5.5

$$\frac{2}{N}\,\text{[birdtrack]} + \text{[birdtrack]} + \text{[birdtrack]} = 4\left(\text{[birdtrack]} + \text{[birdtrack]}\right). \tag{5.13}$$

Observing that the r.h.s. of (5.13) is invariant under a "rotation" of 90°, we deduce a beautiful and quite useful identity:

$$\frac{2}{N}\,\text{[birdtrack]} + \text{[birdtrack]} + \text{[birdtrack]} = \frac{2}{N}\,\text{[birdtrack]} + \text{[birdtrack]} + \text{[birdtrack]}. \tag{5.14}$$

Let us mention that this identity can also be proven as follows. Starting from

$$\left[\left[T^a, T^b\right], T^c\right] = \left\{T^a, \left\{T^b, T^c\right\}\right\} - \left\{T^b, \left\{T^a, T^c\right\}\right\} \tag{5.15}$$

and using (1.4) and (5.5), one arrives at (MacFarlane et al. 1968):

$$f_{abe}\,f_{cde} = \frac{2}{N}\,(\delta_{ac}\delta_{bd} - \delta_{bc}\delta_{ad}) + d_{ace}d_{bde} - d_{ade}d_{bce}. \tag{5.16}$$

Translating into birdtracks, we obtain:

$$\text{[birdtrack]} = \frac{2}{N}\left(\text{[birdtrack]} - \text{[birdtrack]}\right) + \text{[birdtrack]} - \text{[birdtrack]}. \tag{5.17}$$

Check that this is equivalent to Eq. (5.14)!

Exercise 5.6   Multiply (5.14) on the right successively by [birdtrack], [birdtrack], and [birdtrack] to obtain simplifying rules for the following birdtracks:

$$\text{[birdtrack]} \;;\quad \text{[birdtrack]} \;;\quad \text{[birdtrack]}. \tag{5.18}$$

To complement the rules (1.33), let us sum up the new rules involving the star:

$$\text{[birdtrack]} = 0\;;\quad \text{[birdtrack]} = 0$$

$$\text{[birdtrack]} = \frac{N^2 - 4}{N}\,\text{[birdtrack]}\;;\quad \text{[birdtrack]} = \frac{N^2 - 4}{2N}\,\text{[birdtrack]}$$

$$\text{(diagram)} = \frac{N}{2}\ \text{(diagram)}\ ; \quad \text{(diagram)} = \frac{N^2 - 4}{2N}\ \text{(diagram)}$$

$$\text{(diagram)} = \frac{N^2 - 12}{2N}\ \text{(diagram)} \tag{5.19}$$

Exercise 5.7 Show that for $N = 2$, the star fades away.

Exercise 5.8 Evaluate the following graphs in a maximum of three handwritten lines (on an A4 sheet):

$$\text{(diagram)} = ? \ ; \quad \text{(diagram)} = ?$$

5.2   Irreps of a Gluon Pair

We now apply the tensor method to find the irreps of a gluon pair and the associated Hermitian projectors. These projectors have of course been derived in numerous references, see for example Cvitanović (2008), Dokshitzer and Marchesini (2006), MacFarlane et al. (1968) to name but a few.

5.2.1   Basis of Tensors

Using the standard algorithm to find a basis of tensors mapping the gluon pair vector space $\{g^a g^b\} \equiv \mathcal{A} \otimes \mathcal{A}$ to itself, we end up with graphs containing neither three-gluon vertices nor internal gluon lines. Such graphs can be classified as follows:

- *Diagrams with no quark loop*: there are only three of those, namely,

$$\text{(diagram)}\ ; \quad \text{(diagram)}\ ; \quad \text{(diagram)}\ . \tag{5.20}$$

- *Diagrams with a single quark loop*: the diagrams of this type which are not directly proportional to the graphs (5.20) are of the form (diagram), with 3! possible permutations accounting for the different orderings of the gluon vertices around the quark loop. This gives the six tensors:

$$B_+ = \text{(diagram)}\ ; \quad Y_+ = \text{(diagram)}\ ; \quad W_+ = \text{(diagram)}\ ,$$

$$B_- = \text{(diagram)}\ ; \quad Y_- = \text{(diagram)}\ ; \quad W_- = \text{(diagram)}\ . \tag{5.21}$$

- *Diagrams with two or more quark loops*: we can easily check that such diagrams do not provide any new tensors independent of those already found above.

We thus obtain a set of nine tensors, whose linear combinations generate all $SU(N)$ invariant tensors mapping $\mathcal{A} \otimes \mathcal{A} \to \mathcal{A} \otimes \mathcal{A}$. The goal of the next exercise is to show that this set can be traded for the nine tensors given in (5.27) below.

Exercise 5.9 (The identity (5.22) was already proven in Exercise 5.5, but is rewritten here for convenience. Prove the other identities.)

$$4\left(\ \vcenter{\hbox{\includegraphics{}}}\ +\ \vcenter{\hbox{\includegraphics{}}}\ \right) = \vcenter{\hbox{\includegraphics{}}} + \vcenter{\hbox{\includegraphics{}}} + \frac{2}{N}\vcenter{\hbox{\includegraphics{}}} \tag{5.22}$$

$$4\left(\ \vcenter{\hbox{\includegraphics{}}}\ +\ \vcenter{\hbox{\includegraphics{}}}\ \right) = \vcenter{\hbox{\includegraphics{}}} - \vcenter{\hbox{\includegraphics{}}} + \frac{2}{N}\vcenter{\hbox{\includegraphics{}}} \tag{5.23}$$

$$4\left(\ \vcenter{\hbox{\includegraphics{}}}\ -\ \vcenter{\hbox{\includegraphics{}}}\ \right) = \vcenter{\hbox{\includegraphics{}}} + \vcenter{\hbox{\includegraphics{}}} = \vcenter{\hbox{\includegraphics{}}} + \vcenter{\hbox{\includegraphics{}}} \tag{5.24}$$

$$4\left(\ \vcenter{\hbox{\includegraphics{}}}\ -\ \vcenter{\hbox{\includegraphics{}}}\ \right) = \vcenter{\hbox{\includegraphics{}}} - \vcenter{\hbox{\includegraphics{}}} = \vcenter{\hbox{\includegraphics{}}} + \vcenter{\hbox{\includegraphics{}}} \tag{5.25}$$

Inspecting the identities (5.22)–(5.25), we see that the tensors B_+, B_-, Y_+, Y_- of (5.21) can be traded for the tensors

$$\vcenter{\hbox{\includegraphics{}}}\ ;\ \vcenter{\hbox{\includegraphics{}}}\ ;\ \vcenter{\hbox{\includegraphics{}}}\ ;\ \vcenter{\hbox{\includegraphics{}}}\ . \tag{5.26}$$

We can thus choose the following set of nine tensors:

$$I \equiv \vcenter{\hbox{\includegraphics{}}}\ ;\ X \equiv \vcenter{\hbox{\includegraphics{}}}\ ;\ S \equiv \vcenter{\hbox{\includegraphics{}}}\ ;\ F \equiv \vcenter{\hbox{\includegraphics{}}}\ ;\ D \equiv \vcenter{\hbox{\includegraphics{}}}\ ;$$

$$W_+ \equiv \vcenter{\hbox{\includegraphics{}}}\ ;\ W_- \equiv \vcenter{\hbox{\includegraphics{}}}\ ;\ A_1 \equiv \vcenter{\hbox{\includegraphics{}}}\ ;\ A_2 \equiv \vcenter{\hbox{\includegraphics{}}}\ . \tag{5.27}$$

Exercise 5.10 For later use, check the following equations (in less than 20 seconds):

$$4\,(W_+ + W_-) = 4\left(\ \vcenter{\hbox{\includegraphics{}}}\ +\ \vcenter{\hbox{\includegraphics{}}}\ \right) = \vcenter{\hbox{\includegraphics{}}} - \vcenter{\hbox{\includegraphics{}}} + \frac{2}{N}\vcenter{\hbox{\includegraphics{}}}\ , \tag{5.28}$$

$$4 \left(W_+ - W_- \right) = 4 \left(\text{birdtrack} - \text{birdtrack} \right) = \text{birdtrack} - \text{birdtrack} . \tag{5.29}$$

Remember that in order to build the irreps of a gg pair, we need to find a maximal set of linearly independent *commuting* Hermitian tensors mapping $\mathcal{A} \otimes \mathcal{A}$ to itself. The tensors of the set (5.27) are all Hermitian except A_1 and A_2. Since $(A_1)^\dagger = A_2$, one might think of trading A_1 and A_2 for $A_1 \pm A_2$ (as was done for the operators σ and τ in the case of the qqq system, see Sect. 4.3), and since $A_1 + A_2$ is Hermitian, we can obtain eight Hermitian tensors. However, as we will see in the next section, the seven tensors of the subset

$$\mathcal{T} \equiv \{ I, X, S, F, D, W_+, W_- \} \tag{5.30}$$

all commute with each other (see Table 5.1), whereas $A_1 + A_2$ does not commute with some of them. (For instance, one readily checks that $A_1 + A_2$ *anticommutes* with X.) This shows that $\mathcal{T}$ is a good set of tensors to construct gg irreps, which are therefore seven (for $N > 3$, see below the particular case $N = 3$).

Using the same reasoning as in Sect. 4.3 for the qqq system, we infer that A_1 and A_2 (or some linear combinations of them) must correspond to *transition operators* between some equivalent irreps, and the latter must therefore be two. In fact, since A_1 and A_2 are nilpotent,

$$A_1^2 = \text{birdtrack} = 0 ; \quad A_2^2 = \text{birdtrack} = 0 , \tag{5.31}$$

they must correspond precisely to the transition operators. From the pictorial form of A_1 and A_2 (where a single gluon appears in the intermediate state), it directly follows that the two equivalent irreps must be *adjoint* representations. Indeed, we will find the irreps $\mathbf{8}_a$ and $\mathbf{8}_s$ in the color decomposition of a gg pair (see (5.48)), A_1 and A_2 corresponding to the $\mathbf{8}_a \to \mathbf{8}_s$ and $\mathbf{8}_s \to \mathbf{8}_a$ transition operators, respectively.

Strictly speaking, the number of gg irreps is seven if the tensors of $\mathcal{T}$ are linearly independent, which we have not yet proved. For each of the systems considered previously $(qq, q\bar{q}, qg, qqq)$, it was quite obvious that the tensors provided by the standard algorithm were linearly independent, but it is not always the case. For a gluon pair, we will prove *a posteriori* that the tensors of $\mathcal{T}$ are linearly independent when $N > 3$, leading to seven irreps for any $N > 3$, but linearly related when $N = 3$, leading to six irreps in that case (see Exercise 5.12 and following paragraph). This subtlety is however not relevant in the derivation of the irreps and associated projectors. Indeed, the birdtrack identities derived in the present section are clearly valid for $N \geq 3$, and technically there is no need to consider $N = 3$ as a special case.

Table 5.1 Multiplication table of the set of tensors $\mathcal{T} = \{I, X, S, F, D, W_+, W_-\}$. The trivial entries $IX = X$, $IS = S$, ... involving the identity I are not mentioned. We denote $K_A \equiv N^2 - 1$

$\cdot$	X	S	F	D	W_+	W_-
X	I	S	$-F$	D	W_-	W_+
S	S	$K_A S$	0	0	$-\frac{S}{4N}$	$-\frac{S}{4N}$
F	$-F$	0	NF	0	0	0
D	D	0	0	$\frac{N^2-4}{N}D$	$-\frac{D}{2N}$	$-\frac{D}{2N}$
W_+	W_-	$-\frac{S}{4N}$	0	$-\frac{D}{2N}$	$\frac{1}{16}\left[I - \frac{F+D}{N} - \frac{S}{N^2}\right]$	$\frac{1}{16}\left[X + \frac{F-D}{N} - \frac{S}{N^2}\right]$
W_-	W_+	$-\frac{S}{4N}$	0	$-\frac{D}{2N}$	$\frac{1}{16}\left[X + \frac{F-D}{N} - \frac{S}{N^2}\right]$	$\frac{1}{16}\left[I - \frac{F+D}{N} - \frac{S}{N^2}\right]$

5.2.2 Multiplication Table

The multiplication table of the set $\mathcal{T}$ of tensors is presented in Table 5.1. By using the birdtrack expressions of the tensors given in (5.27), the simplest entries of this table can be verified visually, and the other entries obtained by elementary birdtrack calculations. This is left as an exercise.

From Table 5.1, we observe that the tensors of the set $\mathcal{T}$ commute with each other (as already mentioned). Moreover, since the tensors S, F and D have a square proportional to themselves, we directly infer that each of them will provide (after appropriate normalization) the projector on one of the seven gg irreps.

5.2.3 Projectors

With the set of commuting Hermitian tensors and multiplication table in hand, we are ready to derive the projectors on the irreps of a gluon pair. We apply the method already used in Chap. 4 for the qg and qqq systems, which consists in separating the complete space in smaller and smaller invariant subspaces, by considering the successive action of different operators and finding their characteristic equations and eigenspaces. The process is repeated until the number of invariant subspaces equals the number of irreps (which ensures that each subspace cannot be further divided).

First, we observe that $X^2 = I$, implying that X has eigenvalues $\{1, -1\}$. By going to a basis of $\{g^a g^b\} \equiv \mathcal{A} \otimes \mathcal{A}$ where X is diagonal, we find the projectors on the two eigenspaces in invariant form:

$$\mathcal{P}_S = \frac{1}{2}(I + X) = \frac{1}{2}\left(\,\text{⟋⟍} + \text{⟋⟍}\,\right) \;\; ; \;\; \mathcal{P}_A = \frac{1}{2}(I - X) = \frac{1}{2}\left(\,\text{⟋⟍} - \text{⟋⟍}\,\right) .$$

$$(5.32)$$

The projectors (5.32) acting on the digluon space $\mathcal{A} \otimes \mathcal{A}$ are obviously analogous to the diquark symmetrizer and antisymmetrizer (acting on $\mathcal{V} \otimes \mathcal{V}$) defined by (3.2). In the diquark case, we have two irreps, and these projectors are thus sufficient to separate the full space into *irreducible* subspaces. For a gluon pair, we need to find seven irreps, so the projectors $\mathcal{P}_S$ and $\mathcal{P}_A$ only realize a partial decomposition of the complete space into two SU(N) invariant subspaces. The further separation of img($\mathcal{P}_S$) (resp. img($\mathcal{P}_A$)) into irreducible subspaces will provide the so-called symmetric (resp. antisymmetric) irreps of a gg pair.

Antisymmetric irreps

Let us first restrict to img($\mathcal{P}_A$), and use the operators F and W_+ to further separate this subspace. Since $F^2 \mathcal{P}_A = NF\mathcal{P}_A$, the eigenvalues of F are $\{N, 0\}$. In a basis of img($\mathcal{P}_A$) where F is diagonal, we infer the projectors on the corresponding eigenspaces:

$$P_{(N)} = \frac{1}{N} F \mathcal{P}_A = \frac{1}{N} F \equiv \mathbb{P}_a = \mathbb{P}_a \mathcal{P}_A \; ; \quad P_{(0)} = \mathcal{P}_A - \mathbb{P}_a = (\mathcal{P}_A - \mathbb{P}_a)\mathcal{P}_A \; ,$$

(5.33)

where a factor $\mathcal{P}_A$ is highlighted to recall that these projectors vanish on img($\mathcal{P}_S$) and can thus be considered as maps from img($\mathcal{P}_A$) $\rightarrow$ img($\mathcal{P}_A$). (Since all tensors of $\mathcal{T}$ commute, the factor $\mathcal{P}_A$ can be written on the left or right, as desired.)

From the multiplication table, we have $W_+ \mathbb{P}_a \propto W_+ F = 0$, indicating that the subspace img($\mathbb{P}_a$) cannot be further separated by W_+, and therefore that $\mathbb{P}_a$ projects on an irrep. We thus now restrict to img($\mathcal{P}_A - \mathbb{P}_a$), and consider the action of W_+^2 in this subspace:

$$W_+^2 (\mathcal{P}_A - \mathbb{P}_a) = \frac{1}{16}\left[I - \frac{F+D}{N} - \frac{S}{N^2}\right]\mathcal{P}_A(\mathcal{P}_A - \mathbb{P}_a) = \frac{1}{16}(\mathcal{P}_A - \mathbb{P}_a) \; , \quad (5.34)$$

where we used $D\mathcal{P}_A = S\mathcal{P}_A = 0$ and the fact that $\mathcal{P}_A - \mathbb{P}_a$ is a projector. We see that the eigenvalues of W_+ in the subspace img($\mathcal{P}_A - \mathbb{P}_a$) are $\{\frac{1}{4}, -\frac{1}{4}\}$, and the projectors on the eigenspaces read

$$P_{(\pm \frac{1}{4})} = \left(\frac{1}{2} \pm 2W_+\right)(\mathcal{P}_A - \mathbb{P}_a) \; . \quad (5.35)$$

The subspace img($\mathcal{P}_A - \mathbb{P}_a$) cannot be further reduced, and we have thus found the projectors $\mathbb{P}_\alpha$ on three antisymmetric irreps, which we label by $\alpha = \{\mathbf{8}_a, \mathbf{10}, \overline{\mathbf{10}}\}$ according to their dimensions when $N = 3$ (see Exercise 5.12 below):

$$\mathbb{P}_{\mathbf{8}_a} = \mathbb{P}_a = \frac{1}{N} \; \text{⁓⁓⁓} \; , \quad (5.36)$$

$$\mathbb{P}_{\mathbf{10}} = \left(\frac{1}{2} + 2W_+\right)\left(\frac{I - X}{2} - \mathbb{P}_a\right) = \frac{I - X}{4} - \frac{1}{2}\mathbb{P}_a + W_+ - W_- \; , \quad (5.37)$$

$$\mathbb{P}_{\overline{\mathbf{10}}} = \left(\frac{1}{2} - 2W_+\right)\left(\frac{I - X}{2} - \mathbb{P}_a\right) = \frac{I - X}{4} - \frac{1}{2}\mathbb{P}_a - W_+ + W_- \; . \quad (5.38)$$

Symmetric irreps

We now restrict to img($\mathcal{P}_S$), and use the operators S, D and W_+ to separate this subspace. Using $S^2 \mathcal{P}_S = K_A S \mathcal{P}_S$ (with $K_A \equiv N^2 - 1$), the operator S has eigenvalues $\{K_A, 0\}$, with corresponding projectors on eigenspaces:

$$P_{(K_A)} = \frac{1}{K_A} S \mathcal{P}_S = \frac{1}{K_A} S \equiv \mathbb{P}_1 = \mathbb{P}_1 \mathcal{P}_S ; \quad P_{(0)} = \mathcal{P}_S - \mathbb{P}_1 = (\mathcal{P}_S - \mathbb{P}_1)\mathcal{P}_S .$$
(5.39)

The subspace img($\mathbb{P}_1$) has dimension one (see Exercise 5.12), and thus cannot be further reduced. (If we do not see right away that $\mathbb{P}_1$ has rank one, the irreducibility of img($\mathbb{P}_1$) follows from $D\mathbb{P}_1 = 0$ and $W_\pm \mathbb{P}_1 = -\frac{1}{4N}\mathbb{P}_1$, i.e. $D = 0$ and $W_\pm \propto \mathbb{1}$ when restricting to img($\mathbb{P}_1$). Only those operators having at least two eigenvalues in some subspace can be used to separate this subspace.)

In the subspace img($\mathcal{P}_S - \mathbb{P}_1$), the operator D satisfies $D^2 = \frac{N^2-4}{N} D$ (as it actually does in the complete space), and the projectors on its eigenspaces read (for the eigenvalues $\frac{N^2-4}{N}$ and 0, respectively):

$$\frac{N}{N^2-4} D (\mathcal{P}_S - \mathbb{P}_1) = \frac{N}{N^2-4} D \equiv \mathbb{P}_s = \mathbb{P}_s (\mathcal{P}_S - \mathbb{P}_1) ,$$
(5.40)
$$\mathcal{P}_S - \mathbb{P}_1 - \mathbb{P}_s = (\mathcal{P}_S - \mathbb{P}_1 - \mathbb{P}_s)(\mathcal{P}_S - \mathbb{P}_1) .$$
(5.41)

Finally, in the subspace img($\mathcal{P}_S - \mathbb{P}_1 - \mathbb{P}_s$), the operator W_+^2 acts as follows:

$$W_+^2 (\mathcal{P}_S - \mathbb{P}_1 - \mathbb{P}_s) = \frac{1}{16}\left[I - \frac{F+D}{N} - \frac{S}{N^2}\right] (\mathcal{P}_S - \mathbb{P}_1 - \mathbb{P}_s) = \frac{1}{16} (\mathcal{P}_S - \mathbb{P}_1 - \mathbb{P}_s),$$
(5.42)

showing that W_+ has eigenvalues $\{\frac{1}{4}, -\frac{1}{4}\}$ in this subspace, with projectors on eigenspaces:

$$P'_{(\pm\frac{1}{4})} = \left(\frac{1}{2} \pm 2W_+\right)(\mathcal{P}_S - \mathbb{P}_1 - \mathbb{P}_s) .$$
(5.43)

We have thus found four symmetric irreps, labelled by $\alpha = \{\mathbf{1}, \mathbf{8}_s, \mathbf{27}, \mathbf{0}\}$ (see Exercise 5.12), characterized by the projectors

$$\mathbb{P}_1 = \frac{1}{K_A} \text{ }$$
(5.44)

$$\mathbb{P}_{8_s} = \mathbb{P}_s = \frac{N}{N^2-4} \text{ }$$
(5.45)

$$\mathbb{P}_{27} = \left(\frac{1}{2} + 2W_+\right)\left(\frac{I+X}{2} - \mathbb{P}_1 - \mathbb{P}_s\right)$$

$$= \frac{I+X}{4} - \frac{N-2}{2N}\mathbb{P}_s - \frac{N-1}{2N}\mathbb{P}_1 + W_+ + W_- , \tag{5.46}$$

$$\mathbb{P}_0 = \left(\frac{1}{2} - 2W_+\right)\left(\frac{I+X}{2} - \mathbb{P}_1 - \mathbb{P}_s\right)$$

$$= \frac{I+X}{4} - \frac{N+2}{2N}\mathbb{P}_s - \frac{N+1}{2N}\mathbb{P}_1 - W_+ - W_- , \tag{5.47}$$

where the expanded forms of $\mathbb{P}_{27}$ and $\mathbb{P}_0$ easily follow from the multiplication rules of W_+ with the other tensors given in Table 5.1.

In summary, a gluon pair decomposes into a sum of $SU(N)$ irreps as:

$$\mathbf{8} \otimes \mathbf{8} = \mathbf{8}_a \oplus \mathbf{10} \oplus \overline{\mathbf{10}} \oplus \mathbf{1} \oplus \mathbf{8}_s \oplus \mathbf{27} \oplus \mathbf{0} , \tag{5.48}$$

$$\rotatebox{0}{} = \mathbb{P}_{\mathbf{8}_a} + \mathbb{P}_{\mathbf{10}} + \mathbb{P}_{\overline{\mathbf{10}}} + \mathbb{P}_{\mathbf{1}} + \mathbb{P}_{\mathbf{8}_s} + \mathbb{P}_{\mathbf{27}} + \mathbb{P}_{\mathbf{0}} , \tag{5.49}$$

where the first three irreps are antisymmetric, and the last four symmetric, under the permutation of the two incoming (or two outgoing) gluons.

Exercise 5.11 Although the seven projectors $\mathbb{P}_\alpha$, with $\alpha = \{\mathbf{8}_a, \mathbf{10}, \overline{\mathbf{10}}, \mathbf{1}, \mathbf{8}_s, \mathbf{27}, \mathbf{0}\}$, are mutually orthogonal by construction, verify it directly by a simple observation of Eqs. (5.36)–(5.38) and (5.44)–(5.47) (and with a little help from Table 5.1).

Exercise 5.12 Calculate the dimensions of the seven gg irreps as a function of N, and verify that for $N = 3$ these dimensions agree with the names of the irreps.

This last exercise shows that when $N > 3$, the seven projectors are linearly independent (since they are mutually orthogonal and satisfy $\mathrm{rank}(\mathbb{P}_\alpha) \geq 1$). This proves a posteriori that the seven tensors of the set $\mathcal{T}$ (see (5.30)) are also linearly independent when $N > 3$. For $N = 3$ however, $\mathbb{P}_0$ has a null rank, implying $\mathbb{P}_0 = 0$ and a linear relation between the seven tensors. For $N = 3$ only six tensors are independent, and only six irreps remain in the decomposition (5.48)–(5.49).

Exercise 5.13 Check that the projectors $\mathbb{P}_\alpha$ for $\alpha = \{\mathbf{10}, \overline{\mathbf{10}}, \mathbf{27}, \mathbf{0}\}$ can be written as

$$\mathbb{P}_{\mathbf{10}} = 4\left(\mathcal{P}_A - \mathbb{P}_a\right) \qquad\qquad , \tag{5.50}$$

$$\mathbb{P}_{\overline{\mathbf{10}}} = 4\left(\mathcal{P}_A - \mathbb{P}_a\right) \qquad\qquad , \tag{5.51}$$

$$\mathbb{P}_{27} = 4\left(\mathcal{P}_S - \mathbb{P}_1 - \mathbb{P}_s\right) \quad \text{[diagram]} \quad , \tag{5.52}$$

$$\mathbb{P}_0 = 4\left(\mathcal{P}_S - \mathbb{P}_1 - \mathbb{P}_s\right) \quad \text{[diagram]} \quad , \tag{5.53}$$

where the white and black rectangles stand for the symmetrizer and antisymmetrizer over two quark (or antiquark) indices (see (3.2) in Chap. 3). Show that the gg irrep **10** has been correctly named, by explaining why it is equivalent, for $N = 3$, to the qqq irrep **10** identified in Chap. 4. (Use similar reasoning to Exercise 4.7.)

Exercise 5.14 If you want to construct the projectors on the irreps of a system of *three* gluons, try first to find the number of tensors mapping $\{g^a g^b g^c\} \equiv \mathcal{A}^{\otimes 3}$ to itself provided by the standard algorithm. (The answer can be found in Table 1 of Keppeler (2017).)

5.2.4 Some SU(3) Identities

Exercise 5.15 When $N = 3$, show that the linear relation between the seven tensors (5.30), namely, $\mathbb{P}_0^{N=3} = 0$, is equivalent to

$$(N = 3) \quad B_+ + B_- + Y_+ + Y_- + W_+ + W_- = \frac{1}{4}\left(\text{[diagram]} + \text{[diagram]} + \text{[diagram]} \right), \tag{5.54}$$

where the l.h.s. is the sum of the tensors listed in (5.21).

Exercise 5.16 Here is an alternative proof of (5.54). For $N \geq 3$, using (5.47) and the sum of Eqs. (5.22) and (5.23), we readily see that $\mathbb{P}_0$ is a linear combination of $\{I, X, S, B, Y, W\}$, with $B \equiv B_+ + B_-, Y \equiv Y_+ + Y_-, W \equiv W_+ + W_-$. Without using the actual coefficients of this linear combination but only permutations between external gluons, show that the identity $\mathbb{P}_0^{N=3} = 0$ implies a fortiori $B + Y + W = \rho\left(I + X + S\right)$. Then determine the coefficient ρ.

Exercise 5.17 Show the nice SU(3)-relation satisfied by the star d_{abc} tensor (MacFarlane et al. 1968):

$$(N = 3) \quad \text{[diagram]} + \text{[diagram]} + \text{[diagram]} = \frac{1}{3}\left(\text{[diagram]} + \text{[diagram]} + \text{[diagram]} \right). \tag{5.55}$$

5.3   Casimir Charges of gg Irreps

In Chap. 3 we have seen that for a parton pair in irrep R, the Casimir operator is defined by (3.18), and the Casimir charge given by (3.19) in terms of the individual Casimir charges and color interaction potential (3.20). For a gg irrep $\alpha \in \{8_a, 10, \overline{10}, 1, 8_s, 27, 0\}$, we thus have

$$C_\alpha = 2N + V_{gg}(\alpha) , \tag{5.56}$$

where the gg color potential in color state α satisfies:

$$V_{gg}(\alpha)\, \mathbb{P}_\alpha \; = \; -2 \;\; \left(\mathbb{P}_\alpha \right) . \tag{5.57}$$

Since is Hermitian and commutes with each $\mathbb{P}_\alpha$, it must be a linear combination of the projectors:

$$= \sum_\beta a_\beta\, \mathbb{P}_\beta . \tag{5.58}$$

Inserting this in (5.57) we directly get $V_{gg}(\alpha) = -2a_\alpha$. Thus, one just needs the explicit linear combination (5.58) in order to get the color interaction potential for all irreps.

Using (5.14) and (5.28), we can express the l.h.s. of (5.58) as a linear combination of the tensors S, F, D, I, and $W \equiv W_+ + W_-$. Then, W can be traded for a linear combination of projectors by using (5.46) and (5.47). The outcome is (check it!):

$$= N\,\mathbb{P}_1 + \frac{N}{2}\,\mathbb{P}_a + \frac{N}{2}\,\mathbb{P}_s - \mathbb{P}_{27} + \mathbb{P}_0 . \tag{5.59}$$

This gives $V_{gg}(\alpha) = \{-2N, -N, -N, 0, 0, 2, -2\}$ for $\alpha = \{1, 8_a, 8_s, 10, \overline{10}, 27, 0\}$, and the Casimirs C_α follow from (5.56).

5.4   Color States of a Gluon Pair at a Glance

The dimensions K_α, potentials $V_{gg}(\alpha)$ and Casimirs C_α of the gg irreps α are collected in Table 5.2. In certain situations, it can be practical to have this table at hand, and you can legitimately think of tattooing it on a part of the body of your choice.

Table 5.2 Main properties of the SU(N) irreps of a gluon pair. (We denote $K_A \equiv N^2 - 1$)

Irrep α	1	8_a	8_s	10	$\overline{10}$	27	0
K_α	1	K_A	K_A	$K_A \frac{N^2-4}{4}$	$K_A \frac{N^2-4}{4}$	$\frac{N^2(N-1)(N+3)}{4}$	$\frac{N^2(N+1)(N-3)}{4}$
$V_{gg}(\alpha)$	$-2N$	$-N$	$-N$	0	0	2	-2
C_α	0	N	N	$2N$	$2N$	$2(N+1)$	$2(N-1)$

Exercise 5.18 Evaluate in three lines the color graph:

   This birdtrack is given as an example in the Introduction of Cvitanović (2008), where it is evaluated along the lines of the standard algorithm (get rid of three-gluon vertices and then of internal gluon lines by using the Fierz identity). This algorithm is a systematic method of simplifying a color graph, which can be automated using symbolic calculation programs such as FORM (Kuipers et al. 2013). But in some cases, a shortcut can make manual calculation possible and even fast, as is the case here.

Chapter 6
Interlude: Transverse Momentum Broadening in Proton-Nucleus Collisions

Abstract To emphasize that the study of the color structure of parton systems is not just a mathematical game, but aims first and foremost to describe the physical world, we discuss a simple QCD observable which requires knowledge of the irreps and color projectors of parton systems: nuclear transverse momentum broadening of a system of $n \geq 2$ hadronic jets produced in high-energy proton-nucleus collisions.

In previous chapters we have determined the projectors on the color states of any type of parton pair. These projectors are needed to handle certain QCD observables. To take an example (among many others), let's consider the production of two "hadronic jets" in a high-energy proton-proton (pp) collision, as shown schematically in Fig. 6.1. When the transverse momenta K_i of the jets are large, namely, $K_{i\perp} \equiv |K_i| \gg 1\,\mathrm{GeV}$,[1] this process can be described within perturbative QCD (Collins 2011). At leading-order in the strong coupling constant α_s, the two jets arise from a *hard* $2 \to 2$ partonic subprocess, which in Fig. 6.1 is assumed to be $qg \to qg$ (which is only one channel among all possible $2 \to 2$ partonic channels, such as $gg \to gg, gg \to q\bar{q} \ldots$). The $qg \to qg$ scattering amplitude is denoted by

$$\mathcal{M} = \quad \overset{\longrightarrow}{\underset{\text{mm}}{}} \mathcal{M} \overset{\longleftarrow}{\underset{\text{mm}}{}} \, , \tag{6.1}$$

graphically highlighting the initial (incoming) and final (outgoing) partons. For the reader who is not familiar with QCD, note that the above graph does not only represent the color structure of the process. The amplitude $\mathcal{M}$ is given by a sum of *Feynman diagrams*, each diagram being the product of a *color factor* (depending on external parton color indices), and a *Lorentz factor* (depending on kinematical variables as well as on the spins or polarizations of partons). The following discussion is however independent of the details of the amplitude $\mathcal{M}$.

In pp collisions, the two jets are produced nearly "back-to-back" in the transverse plane, i.e. $K_2 \simeq -K_1$. In other words, the pp dijet production cross section

---

[1] In this interlude we use the Planck system of units used in particle physics, where $\hbar = c = 1$.

$d\sigma_{\text{dijet}}/dK_\perp dp_\perp$, where $K_\perp \equiv |\boldsymbol{K}_1 - \boldsymbol{K}_2| \simeq 2K_{1\perp}$ and $p_\perp \equiv |\boldsymbol{K}_1 + \boldsymbol{K}_2|$ are the *relative* and *total* dijet transverse momenta, respectively, is peaked around $p_\perp \simeq 0$. A natural observable in pp collisions is thus $d\sigma_{\text{dijet}}/dK_\perp$, differential in $K_\perp$ but integrated over $p_\perp$. (In the discussion we don't specify jet energies, which are assumed to be fixed and much larger than transverse momenta.) Using QCD factorization theorems (Collins 2011), this observable can be expressed in terms of the parton subprocess cross section $d\hat{\sigma}_{qg \to qg}$, which is itself related to $\sum |\mathcal{M}|^2$ (the sum being over initial and final parton color indices) given by:

$$\sum |\mathcal{M}|^2 = \sum_{i,j,a,b} \overset{i}{\underset{a}{\longrightarrow}} \mathcal{M} \overset{j}{\underset{b}{\longrightarrow}} \overset{j}{\underset{b}{\longrightarrow}} \mathcal{M}^* \overset{i}{\underset{a}{\longrightarrow}} = \mathcal{M} \; \mathcal{M}^*. \quad (6.2)$$

Let's now consider dijet production in proton-nucleus (pA) collisions, by replacing the target proton by a nucleus (represented by the large rectangle in Fig. 6.2) of size L_A. In addition to participating to the $qg \to qg$ subprocess (occurring in an elementary pN collision, where N is a nucleon from the nucleus), the fast incoming quark and outgoing qg pair can now undergo soft rescatterings off the target nucleons encountered along their paths through the nucleus.[2] Since the produced qg pair has a very small transverse size $\sim 1/K_\perp$, it behaves as a pointlike object: soft rescatterings cannot probe its internal structure and thus cannot change $K_\perp$ nor $d\sigma_{\text{dijet}}/dK_\perp$. But rescatterings do see the *global* color charge C_R of the pair (when it is in an irrep R), and can thus deflect it. As a result of the incoming quark and final qg pair rescatterings, in pA collisions the cross section $d\sigma_{\text{dijet}}/dK_\perp dp_\perp$ (at fixed $K_\perp$) is much broader in $p_\perp$ than in pp collisions, which makes it an interesting observable. This effect is quantified by the so-called $p_\perp$-broadening $\Delta p_\perp^2 \equiv \langle p_\perp^2 \rangle_{\text{pA}} - \langle p_\perp^2 \rangle_{\text{pp}} \simeq \langle p_\perp^2 \rangle_{\text{pA}}$.

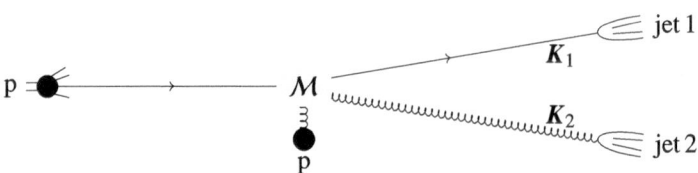

Fig. 6.1 Dijet production in high-energy pp collisions, as viewed in the target proton rest frame. The target proton and right-moving projectile proton are represented by black discs. The incoming quark and gluon entering the $qg \to qg$ hard process (of amplitude $\mathcal{M}$), are picked respectively in the projectile and target protons

---

[2] For simplicity, we consider a limit where the hard subprocess occurs on a time $t_{\text{hard}} \sim E/K_\perp^2 \ll L_A$ (with E the incoming quark energy in the target rest frame). In this limit the blob for $\mathcal{M}$ in Fig. 6.2 is "well located" within the nucleus, allowing to treat independently the hard process and parton propagation before (over the distance L_i) and after (over the distance L_f).

In the above picture we expect $\Delta p_\perp^2$ to receive two contributions:

$$\Delta p_\perp^2\big|_{\text{dijet}} = C_{\text{F}} \cdot \hat{q} L_i + \left(\sum_R \rho_R C_R\right) \cdot \hat{q} L_f, \tag{6.3}$$

where L_i and L_f are the path lengths of the incoming quark and outgoing pointlike qg pair in the nucleus, $\hat{q}$ is a transport coefficient characteristic of the nuclear medium defined by the variation of $p_\perp^2$ per unit length suffered by a fast *unit* color charge ($C_R = 1$), and the sum over R involves the probabilities ρ_R for the qg pair to be produced in irrep R in the elementary pN collision.[3]

In order to determine $\Delta p_\perp^2$, one thus needs to know the irreps R of a qg pair, their Casimirs C_R, and the associated probabilities ρ_R. The projectors $\mathbb{P}_R$ being known, ρ_R is simply obtained by inserting $\mathbb{P}_R$ in the final state of the amplitude $\mathcal{M}$, then squaring and normalizing by $\sum |\mathcal{M}|^2$:

$$\rho_R = \frac{1}{\sum |\mathcal{M}|^2} \; \mathcal{M} \; \boxed{\mathbb{P}_R}\boxed{\mathbb{P}_R^\dagger} \; \mathcal{M}^* = \frac{1}{\sum |\mathcal{M}|^2} \; \mathcal{M} \; \boxed{\mathbb{P}_R} \; \mathcal{M}^*, \tag{6.4}$$

where we used $\mathbb{P}_R^\dagger = \mathbb{P}_R$ and $\mathbb{P}_R^2 = \mathbb{P}_R$. Thus, assuming the production amplitude $\mathcal{M}$ is given, the projectors studied so far are all we need to determine the probabilities (6.4) and predict the $p_\perp$-broadening (6.3). Note that $\sum_R \rho_R = 1$ directly follows from (6.4) by using the completeness relation $\sum_R \mathbb{P}_R = \overrightarrow{}$ and Eq. (6.2).

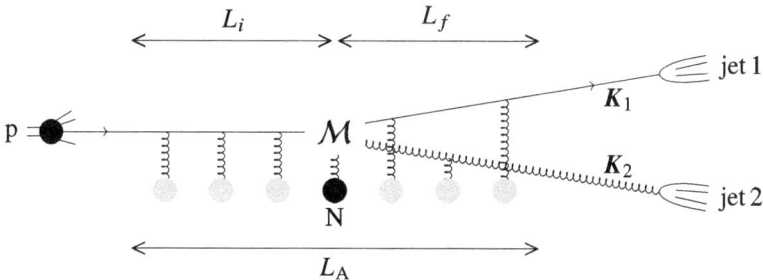

Fig. 6.2 Dijet production in pA collisions. The target nucleus is represented schematically by the large rectangle. The amplitude $\mathcal{M}$ is the same as for pp collisions, but the incoming gluon is now picked in a *nucleon* N (black disc) of the nucleus. Compared to the situation in pp collisions, the fast partons undergo additional scatterings, mediated by soft gluon exchanges with other target nucleons (grey discs)

---

[3] Note that the expression (6.3) of $\Delta p_\perp^2$ as the incoherent sum of the incoming quark and outgoing qg pair $p_\perp$-broadenings, holds only in the limit $t_{\text{hard}} \ll L_A$. When $t_{\text{hard}} \gg L_A$, the hard partonic process is "coherent" over the whole nucleus, and the picture for $p_\perp$-broadening is more complicated (Cougoulic and Peigné 2018).

Interestingly, the projectors $\mathbb{P}_R$ on the irreps of a parton *pair* would also allow to predict $p_\perp$-broadening of a system of *three* jets produced in pA collisions via $2 \to 3$ parton processes. Similarly to dijet production, let's assume the three jets to be produced with large transverse momenta K_i ($i = 1 \ldots 3$), as well as large *relative* momenta $K_{ij} \equiv |K_i - K_j| \sim K_{\text{rel}} \gg 1\,\text{GeV}$, in a specific parton process, for instance $qg \to qgg$ of production amplitude

$$\mathcal{M}' = \overrightarrow{}\, \mathcal{M}' \overrightarrow{}\!\!\! . \tag{6.5}$$

In the limit $t_{\text{hard}} \sim E/K_{\text{rel}}^2 \ll L_{\text{A}}$, the three final partons are produced in a compact configuration within the target nucleus, and $\Delta p_\perp^2$ should be of the same form as (6.3),

$$\Delta p_\perp^2\big|_{3\,\text{jets}} = C_{\text{F}} \cdot \hat{q} L_i + \left(\sum_{R'} \rho_{R'} C_{R'} \right) \cdot \hat{q} L_f , \tag{6.6}$$

where now the sum over R' extends a priori to the available irreps of a qgg system. The probability $\rho_{R'}$ is given by

$$\rho_{R'} = \frac{1}{\sum |\mathcal{M}'|^2} \;\; \mathcal{M}' \; \boxed{\mathbb{P}_{R'}} \; \mathcal{M}'^* , \tag{6.7}$$

with $\mathbb{P}_{R'}$ the projector on the qgg irrep R'.

In (6.6), we can rewrite the average Casimir charge in brackets as

$$\sum_{R'} \rho_{R'} C_{R'} = \frac{1}{\sum |\mathcal{M}'|^2} \sum_R \sum_{R'} C_{R'} \; \boxed{\mathbb{P}_R} \; \mathcal{M}' \; \boxed{\mathbb{P}_{R'}} \; \mathcal{M}'^* , \tag{6.8}$$

where we used the completeness relation for a qg pair, $\overrightarrow{} = \sum_R \mathbb{P}_R$. In the double sum $\sum_R \sum_{R'}$, the irreps R and R' must be equivalent (from Schur's lemma), and thus have the same Casimir. One can thus replace $C_{R'} \to C_R$ in (6.8). Then the only dependence on R' is in $\mathbb{P}_{R'}$, and we can use the completeness relation for a qgg system, $\sum_{R'} \mathbb{P}_{R'} = \overrightarrow{}$, to obtain:

$$\sum_{R'} \rho_{R'} C_{R'} = \frac{1}{\sum |\mathcal{M}'|^2} \sum_R C_R \; \boxed{\mathbb{P}_R} \; \mathcal{M}' \; \overrightarrow{} \mathcal{M}'^* . \tag{6.9}$$

Thus, the $p_\perp$-broadening (6.6) of a three-jet system can in fact be determined by the projectors of a parton pair, as a direct consequence of Schur's lemma. The little game above illustrates the power of Schur's lemma, of which we'll see other examples in subsequent chapters. From the above discussion, it is obvious that parton pair

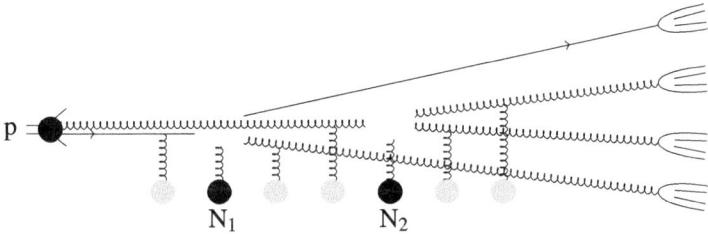

Fig. 6.3 Production of four jets in a high-energy pA collision (viewed in the nucleus rest frame), via two independent $2 \to 2$ parton scatterings (represented by the small grey rectangles), chosen here to be $qg \to qg$ off nucleon N_1, and $gg \to gg$ off nucleon N_2

projectors will be sufficient to predict the $p_\perp$-broadening of systems of $n \geq 3$ jets produced via $2 \to n$ parton processes.

Of course, we should not conclude that parton pair projectors are sufficient in general! Figure 6.3 shows a process that can occur in a pA collision, where *two* incoming partons are picked in the projectile proton, each of them inducing some hard $2 \to 2$ parton scattering off nucleons N_1 and N_2 of the nucleus, leading to a final state containing four hadronic jets. (When $N_1 = N_2$, this process coincides with the four-jet production process in pp collisions studied in Blok et al. (2011).) To predict the $p_\perp$-broadening of a system of three (or four) jets produced via this process, the parton pair projectors derived so far are clearly not sufficient. Indeed, since three of the final partons a fortiori arise from different pairs of initial partons, the above manipulation using Schur's lemma can no longer be used. In this case, one really needs to build Hermitian projectors for systems of $n \geq 3$ partons.

In the next chapters we will derive the color projectors of some three-parton systems, in a pedestrian way, building three-parton irreps recursively by adding one parton to the irreps of a pair. The number of available irreps increases rapidly with the number of partons. For instance, for $N = 3$ a three-gluon system has 29 different irreps, and a four-gluon system 166... and these numbers become 51 and 513, respectively, in the large N limit (see Table 1 of Keppeler (2017))! We will therefore limit ourselves to a few selected cases of three-parton systems.

The chapters that follow should, however, provide some useful tools for constructing projectors on higher-dimensional irreps, as well as a good intuition before investigating advanced techniques for handling the proliferation of color states as the number of partons increases (Cvitanović 2008, Keppeler and Sjodahl 2012, Alcock-Zeilinger and Weigert 2017). In any case, the next chapters should enable you to develop your mastery of the birdtrack technique and continue to impress your friends!

Chapter 7
Irreps of qqq and $qq\bar{q}$ States by Recursive Pairing

Abstract Finding the irreps of a three-parton system abc can be done by first decomposing the parton pair ab into a sum of irreps $(ab)_\alpha$, combining the third parton c with each of these irreps, and decomposing each product $(ab)_\alpha c$ using the tensor method studied in previous chapters. In principle, this simple "recursive pairing" method can be used to construct the irreps and associated projectors of systems with any number of partons. It is illustrated here in the simple cases of qqq and $qq\bar{q}$ systems.

7.1 System qqq

This system has already been studied in Chap. 4 using the tensor method, starting with a set of six independent tensors mapping $\mathcal{V}^{\otimes 3} \to \mathcal{V}^{\otimes 3}$, then selecting four commuting Hermitian tensors, from which we found the linear combinations corresponding to the projectors on four irreps. Here we show how these four projectors can be obtained using recursive pairing, which avoids having to find a complete set of linearly independent maps of $\mathcal{V}^{\otimes 3}$. We'll comment on the advantages of recursive pairing at the end of this section.

So let's start by grouping the first two quarks of the qqq system. This quark pair can be in the SU(N) color states $\mathbf{6}$ or $\bar{\mathbf{3}}$, see (3.5)–(3.6). We then combine each of these states with the third quark. This is illustrated by the pictorial identity:

$$\rightrightarrows = \rightarrow\boxed{}\rightarrow + \rightarrow\blacksquare\rightarrow \, , \tag{7.1}$$

which may be formally written as[1]

$$(\mathbf{3} \otimes \mathbf{3}) \otimes \mathbf{3} = \mathbf{6} \otimes \mathbf{3} \oplus \bar{\mathbf{3}} \otimes \mathbf{3} \, . \tag{7.2}$$

---

[1] Remember that an SU(N) irrep is labelled by the value of its dimension when $N = 3$. In case of ambiguity (when two different SU(N) irreps happen to have equal dimensions for $N = 3$), an irrep is characterized by the expression of its associated projector.

© The Author(s), under exclusive license to Springer Nature Switzerland AG 2024   59
S. Peigné, *Color in QCD*, SpringerBriefs in Physics,
https://doi.org/10.1007/978-3-031-53681-6_7

We will now decompose the systems appearing in the r.h.s. of (7.1), denoted as $(qq)_6 q$ and $(qq)_{\bar{3}} q$, into a sum of irreps, using the tensor method.

7.1.1 Product $(qq)_6 q$

Now that we're experts in the tensor method, let's go a little faster than in previous chapters. The first term in the r.h.s. of (7.1) is the identity map of $\{\mu_S^{ij} q^k\}$, where μ_S^{ij} is the symmetric diquark multiplet defined in (3.3). In order to find the irreps of $(qq)_6 q \sim \mathbf{6} \otimes \mathbf{3}$, we first need a basis of tensors mapping $\{\mu_S^{ij} q^k\}$ to itself. These tensors are of the form

$$\rightarrow\!\boxed{\;?\;}\!\rightarrow \tag{7.3}$$

where the blob may a priori account for any possible color graph. Using the standard algorithm (see Chap. 3), we can get rid of gluons in the blob, and we readily find that there are only two independent ways to connect the quark lines on the left to the quark lines on the right. This gives the set of two tensors:

$$I = \rightarrow\!\boxed{\;\;}\!\rightarrow \quad ; \quad A = \rightarrow\!\boxed{\;\;}\!\rightarrow . \tag{7.4}$$

Exercise 7.1 Show that $A^2 = \frac{1}{2}(I + A)$, and that the projectors on the eigenspaces of A associated to the eigenvalues 1 and $-\frac{1}{2}$ read:

$$P_{(1)} \equiv \mathbb{P}_{10} = \frac{1}{3}\,\rightarrow\!\boxed{\;\;}\!\rightarrow + \frac{2}{3}\,\rightarrow\!\boxed{\;\;}\!\rightarrow = \rightarrow\!\boxed{\;\;}\!\rightarrow , \tag{7.5}$$

$$P_{(-1/2)} \equiv \mathbb{P}_{8_+} = \frac{2}{3}\,\rightarrow\!\boxed{\;\;}\!\rightarrow - \frac{2}{3}\,\rightarrow\!\boxed{\;\;}\!\rightarrow = \frac{4}{3}\,\rightarrow\!\boxed{\;\;}\!\rightarrow . \tag{7.6}$$

We recover the projectors $\mathbb{P}_{10}$ and $\mathbb{P}_{8_+}$ found previously in Chap. 4, see Eqs. (4.29) and (4.31), corresponding to the $SU(N)$ irreps $\mathbf{10}$ and $\mathbf{8}_+$ and satisfying the completeness relation $I = \mathbb{P}_{10} + \mathbb{P}_{8_+}$. (The dimensions of the irreps $\mathbf{10}$ and $\mathbf{8}_+$ for general N were calculated in Exercise 4.10.) The system $(qq)_6 q$ thus decomposes as

$$\mathbf{6} \otimes \mathbf{3} = \mathbf{10} \oplus \mathbf{8}_+ . \tag{7.7}$$

Pictorially, the fact that the relation $I = \mathbb{P}_{10} + \mathbb{P}_{8_+}$ expresses the decomposition into irreps of the system $(qq)_6 q$ is obvious: the Hermitian projectors (7.5) and (7.6) are maps of $\{\mu_S^{ij} q^k\}$ (due to the presence of the symmetrizers on the left and right), they are mutually orthogonal (by construction), and we know from the number of independent commuting Hermitian tensors that there must be two irreps.

As already mentioned in the beginning of Chap. 3, an irreducible $SU(N)$ invariant subspace (or irrep) of a system of quarks $\{q^i q^j \dots q^p\}$ is spanned by a set (called multiplet) of linear combinations of $q^i q^j \dots q^p$ characterized by specific symmetry properties in the permutation of indices. In the present case of the three-quark system $\mu_S^{ij} q^k$, the first factor μ_S^{ij} is already symmetrized in ij, and we thus get two possibilities: either a total symmetry in ijk, or a mixed symmetry, corresponding respectively to the irreps **10** and $\mathbf{8}_+$ of associated projectors (7.5) and (7.6).

Exercise 7.2 Check that the multiplets corresponding to the irreps **10** and $\mathbf{8}_+$ are given by

$$[\mu_{\mathbf{10}}]^{ijk} = \frac{1}{3} \left(\mu_S^{ij} q^k + \mu_S^{ik} q^j + \mu_S^{jk} q^i \right) , \tag{7.8}$$

$$[\mu_{\mathbf{8}_+}]^{ijk} = \frac{1}{3} \left(2\mu_S^{ij} q^k - \mu_S^{ik} q^j - \mu_S^{jk} q^i \right) . \tag{7.9}$$

(Remember the definition (2.32) of a multiplet.)

A few remarks:

- Since $\mathbb{P}_{\mathbf{8}_+}$ is a map of $\{\mu_S^{ij} q^k\}$, the corresponding multiplet is symmetric in ij, and therefore spans a subspace of $\{\mu_S^{ij} q^k\}$, as it should. This is why the multiplet (7.9) can be obtained from $\mu_S^{ij} q^k$ by first antisymmetrizing in jk and lastly symmetrizing in ij, as suggested by the pictorial form (7.6) of $\mathbb{P}_{\mathbf{8}_+}$.
- In general, the explicit form of multiplets is not really necessary, since all the information about irreps is contained in the projectors. In what follows, we'll use multiplets for teaching purposes only, to illustrate that projectors obey the rules for characterizing an irrep, as specified by the "index method" (see Chap. 8).
- Beware of a possible error (seen on TV!): the operator $A^2 = \frac{1}{2}(I + A)$ can be expressed pictorially as

$$A^2 = \quad , \tag{7.10}$$

where only diquark symmetrizers appear. A viewer might be tempted to conclude that this operator is totally symmetric in the permutation of quark indices, which would be a big mistake. In fact, A^2 has neither total symmetry nor mixed symmetry under permutation, and therefore cannot be proportional to a projector on an irrep.

7.1.2 Product $(qq)_{\bar{3}}q$

The system $(qq)_{\bar{3}}q \sim \bar{\mathbf{3}} \otimes \mathbf{3}$ (the second term in the r.h.s. of (7.1)), can be studied along the same lines as the system $(qq)_6 q$. For a basis of tensors mapping the space $\{\mu_A^{ij} q^k\}$ to itself (with μ_A^{ij} the antisymmetric diquark multiplet (3.4)), we may choose:

$$I' = \text{[diagram]} \quad ; \quad A' = \text{[diagram]} . \tag{7.11}$$

Exercise 7.3 Show that $(qq)_{\bar{3}}q$ decomposes as

$$\bar{3} \otimes 3 = 1 \oplus 8_- , \tag{7.12}$$

with the projectors associated to the irreps **1** and **8**$_-$ given by:

$$\mathbb{P}_1 = \frac{1}{3}\text{[diagram]} - \frac{2}{3}\text{[diagram]} = \text{[diagram]} , \tag{7.13}$$

$$\mathbb{P}_{8_-} = \frac{2}{3}\text{[diagram]} + \frac{2}{3}\text{[diagram]} = \frac{4}{3}\text{[diagram]} . \tag{7.14}$$

These projectors coincide with the projectors (4.30) and (4.32) found in Chap. 4, and satisfy the completeness relation $I' = \mathbb{P}_1 + \mathbb{P}_{8_-}$ in the space $\{\mu_A^{ij}q^k\}$.

Exercise 7.4 Similarly to Exercise 7.2, write the expressions of the multiplets $[\mu_1]^{ijk}$ and $[\mu_{8_-}]^{ijk}$ associated to the irreps **1** and **8**$_-$.

7.1.3 Sum Up: Decomposition of qqq

To sum up, using (7.1), (7.5)–(7.6), and (7.13)–(7.14), we recover the decomposition of the qqq system into a sum of irreps found in Chap. 4 (see (4.33)):

$$\text{[diagram]} = \underbrace{\frac{1}{3}\text{[diagram]} + \frac{2}{3}\text{[diagram]}}_{\mathbb{P}_{10}} + \underbrace{\frac{2}{3}\text{[diagram]} - \frac{2}{3}\text{[diagram]}}_{\mathbb{P}_{8_+}}$$

$$+ \underbrace{\frac{2}{3}\text{[diagram]} + \frac{2}{3}\text{[diagram]}}_{\mathbb{P}_{8_-}} + \underbrace{\frac{1}{3}\text{[diagram]} - \frac{2}{3}\text{[diagram]}}_{\mathbb{P}_1} . \tag{7.15}$$

Pictorially, it is obvious that this decomposition has been obtained by first pairing quarks 1 and 2, in either the irrep **6** or the irrep $\bar{3}$. The first line corresponds to the further decomposition of $(qq)_6 q$, and the second line to that of $(qq)_{\bar{3}}q$. In each line, projectors are written in such a way that the completeness relations in the respective subspaces $\{\mu_S^{ij}q^k\}$ and $\{\mu_A^{ij}q^k\}$ are obvious.

By comparing with the study of the qqq system in Chap. 4, we can see that recursive pairing avoids having to consider non-Hermitian operators, and provides the four Hermitian projectors more directly. Indeed, for each product of two irreps encountered (either qq in the first step, or $(qq)_6 q$ and $(qq)_{\bar{3}} q$ in the second), the independent tensors are all Hermitian and commute, and their number is thus equal to the number of irreps in the decomposition of the product. The equivalent irreps $\mathbf{8}_+$ and $\mathbf{8}_-$ pose no difficulty, since they appear in the separate products $(qq)_6 q$ and $(qq)_{\bar{3}} q$.

In general, the product of any two SU(N) irreps is likely to contain equivalent irreps.[2] However, the example of the qqq system shows that to find the decomposition of a multiparton system, recursive pairing is more efficient than starting from the complete multiparton vector space. And if we're interested in the transition operators between equivalent irreps, these operators can be easily inferred from the knowledge of Hermitian projectors.

Remember that the equivalent irreps $\mathbf{8}_+$ and $\mathbf{8}_-$ are not uniquely defined separately. With the tensor method used in Chap. 4, this follows from the freedom in the choice of the fourth commuting Hermitian tensor (to form a basis of six independent tensors mapping $\mathcal{V}^{\otimes 3} \to \mathcal{V}^{\otimes 3}$), see Exercise 4.15. Using recursive pairing, this can be explained more naturally by the freedom of choice of the first pair of partons. By choosing to group quarks 1 and 3 first instead of quarks 1 and 2, one would get the same relation as (7.15), up to the permutation of quarks 2 and 3, obtained pictorially by multiplying (7.15) to the left and right by $X_1 = \rightthreetimes$. Since $\mathbb{P}_{10}$ and $\mathbb{P}_1$ (and of course the identity of $\mathcal{V}^{\otimes 3}$) are invariant under this operation, one would thus obtain:

$$\begin{array}{c} \overrightarrow{\quad} \\ \overrightarrow{\quad} \\ \overrightarrow{\quad} \end{array} = \mathbb{P}_{10} + \mathbb{P}_{8'_+} + \mathbb{P}_{8'_-} + \mathbb{P}_1 , \qquad (7.16)$$

with $\mathbb{P}_{8'_+} = X_1 \mathbb{P}_{8_+} X_1$ and $\mathbb{P}_{8'_-} = X_1 \mathbb{P}_{8_-} X_1$ the projectors on the irreps $\mathbf{8}'_+$ and $\mathbf{8}'_-$ (see Exercise 4.15) related to $\mathbf{8}_+$ and $\mathbf{8}_-$ by a change of basis with similarity matrix X_1.

This shows that although the qqq multiplets corresponding to the equivalent irreps $\mathbf{8}_+$ and $\mathbf{8}_-$ do not mix under SU(N), they have no real physical significance individually. This is quite understandable, since the qqq irreps shouldn't care about the order of pairing used to find them. What is physical, however, is that the subspace orthogonal to $\mathbf{1} \oplus \mathbf{10}$ can be split into two irreducible subspaces sharing the same SU(N) invariants.

---

[2] The simplest example of a product of two irreps containing equivalent irreps is $\mathbf{8} \otimes \mathbf{8}$ (which contains two adjoint representations, see the decomposition (5.48) of a gg pair derived in Chap. 5). Thus, each time we'll have to combine a multiparton state in an adjoint representation with a gluon (at some step of the recursive pairing), some non-Hermitian operators will inevitably enter the set of independent tensors. We'll see another example of this in Chap. 9 with the product $\mathbf{8} \otimes \mathbf{15}$, which also contains equivalent irreps.

7.2 System $qq\bar{q}$

We now apply the recursive pairing method to the $qq\bar{q}$ system, to find a decomposition into irreps of the type:

$$\text{(diagram)} = \sum_{\alpha} \text{(diagram } \mathbb{P}_\alpha\text{)} . \tag{7.17}$$

Pairing first the two quarks, we obtain:

$$\text{(diagram)} = \text{(diagram)} + \text{(diagram)} , \tag{7.18}$$

$$(\mathbf{3} \otimes \mathbf{3}) \otimes \bar{\mathbf{3}} = \mathbf{6} \otimes \bar{\mathbf{3}} \oplus \bar{\mathbf{3}} \otimes \bar{\mathbf{3}} . \tag{7.19}$$

Note that the two irreps named $\bar{\mathbf{3}}$ in the product $\bar{\mathbf{3}} \otimes \bar{\mathbf{3}}$ are not the same irreps for $N > 3$, as is pictorially clear (see footnote 1). The antisymmetric diquark irrep has dimension $N(N-1)/2$, whereas the antiquark irrep has dimension N, and these irreps can thus coincide (strictly speaking, be equivalent) only for $N = 3$ (see Exercise 3.3).

In the same way as in the previous section, we can now decompose the systems $(qq)_6\bar{q}$ and $(qq)_{\bar{3}}\bar{q}$. This is done in the following two exercises.

Exercise 7.5 To decompose $(qq)_6\bar{q}$, we choose the following basis of tensors:

$$\text{(diagram)} \quad ; \quad \text{(diagram)} . \tag{7.20}$$

Show that $\mathbf{6} \otimes \bar{\mathbf{3}} = \mathbf{3}_s \oplus \mathbf{15}$, with the irreps $\mathbf{3}_s$ and $\mathbf{15}$ defined by

$$\mathbb{P}_{3_s} = \frac{2}{N+1} \text{(diagram)} , \tag{7.21}$$

$$\mathbb{P}_{15} = \text{(diagram)} - \frac{2}{N+1} \text{(diagram)} . \tag{7.22}$$

What are the dimensions of $\mathbf{3}_s$ and $\mathbf{15}$ for general N?

Exercise 7.6 The system $(qq)_{\bar{3}}\bar{q}$ decomposes as $\bar{\mathbf{3}} \otimes \bar{\mathbf{3}} = \mathbf{3}_a \oplus \bar{\mathbf{6}}$, with

$$\mathbb{P}_{3_a} = \frac{2}{N-1} \text{(diagram)} , \tag{7.23}$$

$$\mathbb{P}_{\bar{6}} = \text{(diagram)} - \frac{2}{N-1} \text{(diagram)} . \tag{7.24}$$

What are the dimensions of $\mathbf{3}_a$ and $\bar{\mathbf{6}}$ for general N?

In summary, using (7.18), (7.21)–(7.22) and (7.23)–(7.24), a $qq\bar{q}$ system decomposes as follows:

$$\overrightarrow{\underset{\longleftarrow}{\longrightarrow}} = \mathbb{P}_{3_s} + \mathbb{P}_{15} + \mathbb{P}_{3_a} + \mathbb{P}_{\bar{6}} \ , \tag{7.25}$$

$$\mathbf{3} \otimes \mathbf{3} \otimes \bar{\mathbf{3}} = \mathbf{3}_s \oplus \mathbf{15} \oplus \mathbf{3}_a \oplus \bar{\mathbf{6}} \ . \tag{7.26}$$

Quite obviously, the irreps $\mathbf{3}_s$ and $\mathbf{3}_a$ are equivalent for any $N \geq 3$. Pictorially, we see that the projectors $\mathbb{P}_{3_s}$ and $\mathbb{P}_{3_a}$ have an intermediate state consisting of a single quark, so from Schur's lemma $\mathbf{3}_s$ and $\mathbf{3}_a$ must be equivalent to the fundamental representation. They are therefore rightly named $\mathbf{3}$, the subscripts s and a simply reminding us in which diquark subspace (symmetric or antisymmetric) the irrep is found in the recursive derivation.

Exercise 7.7 According to the definition (2.32), the multiplet associated to a $qq\bar{q}$ irrep α reads

$$[\mu_\alpha]_k^{ij} \equiv \ \substack{\circ \rightarrow \\ \circ \rightarrow} \left(\mathbb{P}_\alpha \right) \substack{\rightarrow i \\ \rightarrow j \\ \leftarrow k} \ . \tag{7.27}$$

Obtain the following expressions:

$$[\mu_{3_s}]_k^{ij} = \frac{1}{N+1} \left(\mu_S^{il} q_l \delta_k^j + \mu_S^{jl} q_l \delta_k^i \right) \ , \tag{7.28}$$

$$[\mu_{15}]_k^{ij} = \mu_S^{ij} q_k - [\mu_{3_s}]_k^{ij} \ , \tag{7.29}$$

$$[\mu_{3_a}]_k^{ij} = \frac{1}{N-1} \left(\mu_A^{il} q_l \delta_k^j - \mu_A^{jl} q_l \delta_k^i \right) \ , \tag{7.30}$$

$$[\mu_{\bar{6}}]_k^{ij} = \mu_A^{ij} q_k - [\mu_{3_a}]_k^{ij} \ . \tag{7.31}$$

As was the case for the qqq system discussed in Sect. 7.1, recursive pairing allows one to find the $qq\bar{q}$ irreps without having to list all tensors mapping $qq\bar{q} \rightarrow qq\bar{q}$, thus avoiding the discussion of transition operators between the equivalent irreps $\mathbf{3}_s$ and $\mathbf{3}_a$. If needed, the latter operators are not difficult to find.

Exercise 7.8 Find an explicit birdtrack expression of the $\mathbf{3}_s \rightarrow \mathbf{3}_a$ transition operator. Remember that the latter is uniquely defined (up to an overall factor) and must be of the form (see (3.14))

$$T = \ \substack{\rightarrow \\ \rightarrow \\ \leftarrow} \left(\mathbf{3}_s \right) \substack{\rightarrow \\ \rightarrow \\ \leftarrow} \left(\tilde{T} \right) \substack{\rightarrow \\ \rightarrow \\ \leftarrow} \left(\mathbf{3}_a \right) \substack{\rightarrow \\ \rightarrow \\ \leftarrow} \ . \tag{7.32}$$

Thus, all we need to do is find a tensor $\tilde{T}$ that gives a non-zero result when inserted into (7.32). (See the discussion of transition operators for the qqq system in Sect. 4.3.2.)

Chapter 8
SU(N) Irreps from the Index Method

Abstract We present the "index method", based on the characterization of an irrep by quark and antiquark indices and their symmetry properties under permutation, which is a useful auxiliary tool for decomposing the product of any two irreps.

8.1 The Index Method

An irrep of a system of m quarks is associated to a set (or multiplet) of linear combinations of $q^{i_1} \ldots q^{i_m}$ having a specific symmetry in the permutation of quark indices. We have seen examples of this with the qq and qqq systems. A similar statement obviously holds for systems $q_{k_1} \ldots q_{k_n}$ of n antiquarks. It is not difficult to show that any SU(N) irrep can be characterized by a tensor $U^{i_1 i_2 \ldots i_m}_{k_1 k_2 \ldots k_n}$ carrying m upper (quark) and n lower (antiquark) indices, with specific symmetries both in the permutation of upper indices and in the permutation of lower indices, and being furthermore *traceless*: the contraction[1] of any upper index with any lower index gives zero, e.g., $U^{i_1 i_2 i_3 \ldots}_{i_2 k_2 k_3 \ldots} = 0$.

As a consequence, the product of two SU(N) irreps characterized in this way,

$$U^{i_1 i_2 \ldots i_m}_{k_1 k_2 \ldots k_n} \ V^{j_1 j_2 \ldots j_p}_{l_1 l_2 \ldots l_q} \ , \tag{8.1}$$

can be decomposed into a sum of irreps using the following procedure, which we will call the "index method" for short:

---

[1] We may also say 'trace', although such trace should not be confused with the trace over initial and final color indices of a tensor viewed as a map of the parton system vector space.

S. Peigné, *Color in QCD*, SpringerBriefs in Physics,
https://doi.org/10.1007/978-3-031-53681-6_8

1. Find all distinct ways to 'trace' over upper and lower indices. (Only the contractions between the sets of indices $i_1 i_2 \ldots i_m$ and $l_1 l_2 \ldots l_q$, or between the sets $k_1 k_2 \ldots k_n$ and $j_1 j_2 \ldots j_p$, need to be considered, since the tensors U and V are traceless by assumption.)
2. For a given contraction of indices (including the case where there is none), find all possible symmetries (including mixed symmetries) under permutation of the remaining upper, and separately lower, indices.
3. Make each of the tensors obtained after steps 1 and 2 traceless by appropriate subtractions.

The traceless tensors resulting from this procedure are in one-to-one correspondence with the irreps appearing in the decomposition of the product (8.1).

Although the index method does not directly give Hermitian projectors, it is a very helpful tool for calculating them and therefore deserves to be known. In particular, for systems with a small number of partons, the index method gives the number of irreps appearing in the product of two irreps, the symmetries (under permutation of quark or antiquark indices) and the dimensions of these irreps, with minimum effort.

To illustrate how it works, let's return to the derivation of $qq\bar{q}$ irreps of the previous chapter, still using recursive pairing, but using also the index method as a guide.

8.2 Irreps of $qq\bar{q}$ from the Index Method

As in Sect. 7.2, let's first group the quark pair $q^i q^j$. According to the index method, this decomposes into the multiplets μ_S^{ij} and μ_A^{ij}, as we know. (Since $q^i q^j$ contains no antiquark, in this simple case the index method boils down to step 2, i.e. finding all possible symmetries in the permutation of two quark indices.)

We now combine the quark pair with the antiquark, and use the index method to decompose the products $\mu_S^{ij} q_k \sim \mathbf{6} \otimes \bar{\mathbf{3}}$ and $\mu_A^{ij} q_k \sim \bar{\mathbf{3}} \otimes \bar{\mathbf{3}}$.

8.2.1 $\mathbf{6} \otimes \bar{\mathbf{3}}$

At step 1 of the index method, we can take either one trace or none in the product $\mu_S^{ij} q_k$, yielding two objects:

$$\mu_S^{ij} q_k \xrightarrow[\text{one trace}]{\text{step 1}} \mu_S^{il} q_l \ , \tag{8.2}$$

$$\mu_S^{ij} q_k \xrightarrow[\text{zero trace}]{\text{step 1}} \mu_S^{ij} q_k \ . \tag{8.3}$$

To proceed with steps 2 and 3, it is more convenient to start with the objects corresponding to the largest number of traces (which will provide the irreps of lower dimensions in the final decomposition). So let's first consider $\mu_S^{il} q_l$, which carries only one free quark index. Steps 2 and 3 are not applicable, and the quantity (8.2) must therefore provide one irrep, which is clearly equivalent to the fundamental representation. To obtain the specific linear combinations of the $\mu_S^{ij} q_k$'s (i.e. the multiplet) corresponding to this irrep, one needs the associated Hermitian projector, which must be a map of the space $\{\mu_S^{ij} q_k\}$. Pictorially, it should be obvious that this projector is of the form

$$P_3 = c \cdot \quad , \qquad (8.4)$$

where the normalization factor $c = 2/(N+1)$ can easily be found from the condition $P_3^2 = P_3$. We thus recover the projector $\mathbb{P}_{3_s}$ (see (7.21)) and the corresponding multiplet μ_{3_s} given in (7.28).

Note that the expression of this multiplet can be obtained by simple reasoning, without the aid of birdtracks. Since $[\mu_{3_s}]_k^{ij}$ must span a subspace of $\{\mu_S^{ij} q_k\}$, we first multiply $\mu_S^{il} q_l$ by δ_k^j (the only available SU(N) invariant tensor carrying the free indices j and k) and then symmetrize over ij. The normalization factor can be determined by the condition $[\mu_{3_s}]_l^{il} = \mu_S^{il} q_l$, which follows from the fact that the multiplet μ_{15} discussed below (see (8.5)) is traceless.

To avoid any confusion, let us stress that the ij symmetry of the multiplet $[\mu_{3_s}]_k^{ij}$ is not inherent to the irrep $\mathbf{3}_s$, but simply reminds us that this irrep is found in the product $\mu_S^{ij} q_k$. Let us repeat that the irrep $\mathbf{3}_s$ is characterized, from the point of view of SU(N) transformations, by a *single* upper index. This is visible on the expression of μ_{3_s} in (7.28) (the Kronecker's being SU(N) invariant), or pictorially by observing the intermediate state of the projector (8.4).

Let us now discuss the object (8.3) obtained (without much effort) at the first step of the index method. The quantity $\mu_S^{ij} q_k$ has two free quark indices, and one free antiquark index. Since the symmetry in quark indices is already specified, step 2 is completed, and $\mu_S^{ij} q_k$ will thus provide one irrep, but only after it is made traceless (step 3). This is done by subtracting the multiplet $[\mu_{3_s}]_k^{ij}$ from $\mu_S^{ij} q_k$, which yields the multiplet μ_{15}:

$$[\mu_{15}]_k^{ij} = \mu_S^{ij} q_k - [\mu_{3_s}]_k^{ij} = \quad \begin{array}{c} i \\ j \\ k \end{array} - \quad \boxed{3_s} \begin{array}{c} i \\ j \\ k \end{array} . \qquad (8.5)$$

(We recall the obvious notation introduced in Sect. 3.2, whereby a blob denoted by α represents the projector on the irrep α of the partonic system entering or leaving the blob.) We thus recover the multiplet μ_{15} found previously in (7.29), and it trivially follows from (8.5) that the associated projector $\mathbb{P}_{15}$ is as given in (7.22).

The above derivation of the irreps (or equivalently multiplets) of the system $\mu_S^{ij} q_k$, performed using the index method, illustrates some important general features:

- In a subspace (such as $\{\mu_S^{ij} q_k\}$) with specific symmetries in quark and antiquark indices, making a multiplet traceless amounts to remove the projectors already found in this subspace (corresponding to multiplets associated with a larger number of traces) from the identity operator in this subspace.
- The dimension of an irrep can in principle be determined without knowing the exact form of the multiplet, simply by observing the indices and remembering that a multiplet must be traceless. For example, the dimension of the irrep **15** appearing in the decomposition of $\mu_S^{ij} q_k$ can be obtained as follows:

$$\dim\left\{[\mu_{15}]_k^{ij}\right\} = \dim\left\{\left[\mu_S^{ij} q_k\right]_{\text{traceless}}\right\} = \frac{N(N+1)}{2}N - N. \qquad (8.6)$$

In the r.h.s. of (8.6), the first term is the dimension of $\{\mu_S^{ij} q_k\} \sim \mathbf{6} \otimes \bar{\mathbf{3}}$, and the second term accounts for the number of constraints provided by the tracelessness condition $[\mu_{15}]_l^{il} = 0$. This number obviously coincides with the dimension of the irrep $\mathbf{3}_s$ characterized by a single quark index and obtained in (8.2) by taking one trace over upper and lower indices.

8.2.2 $\bar{\mathbf{3}} \otimes \bar{\mathbf{3}}$

The product $\mu_A^{ij} q_k$ can be decomposed following exactly the same lines as for $\mu_S^{ij} q_k$. In brief:

$$\mu_A^{ij} q_k \xrightarrow[\text{one trace}]{\text{step 1}} \mu_A^{il} q_l \;:\; \text{irrep } \mathbf{3}_a \;, \qquad (8.7)$$

$$\mu_A^{ij} q_k \xrightarrow[\text{zero trace}]{\text{step 1}} \mu_A^{ij} q_k \xrightarrow{\text{step 3}} \left[\mu_A^{ij} q_k\right]_{\text{traceless}} \;:\; \text{irrep } \bar{\mathbf{6}} \;, \qquad (8.8)$$

where only those steps of the index method that are applicable are mentioned.

Exercise 8.1 Using the procedure summarized in (8.7) and (8.8), explain how to find the multiplets (7.30) and (7.31) directly.

8.3   Some Training Exercises

To get used to the index method, here are a few exercises, illustrating that this method can be a valuable aid in "clearing the ground" before attacking the calculation of Hermitian projectors on the irreps of a multiparton system.

Exercise 8.2 Recover the irreps of a $q\bar{q}$ pair (found in Sect. 4.1), their associated projectors and multiplets, by applying the index method to the product $q^i q_j$. (See also Exercise 4.1.)

Exercise 8.3 By reasoning only on indices, briefly explain why the $qq\bar{q}$ irrep **15** found above (see (8.5)) is equivalent to the irrep **15** of a qg pair derived in Chap. 4 (of associated projector given by Eq. (4.17)).

Exercise 8.4 Denoting by φ^{ijk} and ψ_{lmn} the totally symmetric color states of three quarks and three antiquarks, respectively, use the index method to show that $\varphi^{ijk}\psi_{lmn}$ decomposes as

$$\mathbf{10} \otimes \overline{\mathbf{10}} = \mathbf{1} \oplus \mathbf{8} \oplus \mathbf{27} \oplus \mathbf{64} . \tag{8.9}$$

Calculate the dimensions of the $SU(N)$ irreps appearing in the latter decomposition as a function of N. (Don't look for projectors!)

Exercise 8.5 From Exercise 8.2, the adjoint representation of $SU(N)$ is characterized by the traceless tensor

$$G^i_j \equiv \ \begin{array}{c}\bigcirc\!\!\rightarrow\\\bigcirc\!\!\leftarrow\end{array}\!\!\left(\mathbf{8}\right)\!\!\begin{array}{c}\rightarrow i\\\leftarrow j\end{array} , \tag{8.10}$$

where the blob **8** denotes the "octet" $q\bar{q}$ projector (4.4). Use the index method to recover the decomposition (5.48) of a gg pair into irreps and the dimensions of those irreps.

8.4 $qq\bar{q}$ Irreps from a Different Pairing Order

As an extra workout, let's see how the $qq\bar{q}$ irreps would arise if we would have chosen to first group together a $q\bar{q}$ pair. Using the $q\bar{q}$ projectors (4.3) and (4.4) and the decomposition (4.5) we have

$$\begin{array}{c}\rightarrow\\\rightarrow\\\leftarrow\end{array} = \begin{array}{c}\rightarrow\\\rightarrow\left(\mathbf{1}\right)\rightarrow\\\leftarrow\end{array} + \begin{array}{c}\rightarrow\\\rightarrow\left(\mathbf{8}\right)\rightarrow\\\leftarrow\end{array} , \tag{8.11}$$

$$\mathbf{3} \otimes (\mathbf{3} \otimes \bar{\mathbf{3}}) = \mathbf{3} \otimes \mathbf{1} \oplus \mathbf{3} \otimes \mathbf{8} . \tag{8.12}$$

The first term in the r.h.s. of (8.11) already gives one $qq\bar{q}$ irrep, obviously equivalent to the fundamental representation, defined by the projector

$$\mathbb{P}_{3'} = \begin{array}{c}\rightarrow\\\rightarrow\left(\mathbf{1}\right)\rightarrow\\\leftarrow\end{array} = \frac{1}{N} \begin{array}{c}\rightarrow\\\,\rfloor\ \rfloor\\\leftarrow\end{array} . \tag{8.13}$$

We refer to this irrep as $\mathbf{3}'$ to distinguish it from the equivalent irreps $\mathbf{3}_s$ and $\mathbf{3}_a$ found in Sect. 7.2 (or in Sect. 8.2) by grouping first the two quarks.

We now decompose the second term of (8.11) into irreps. This can be done directly by using the pictorial expression (4.4) of the octet $q\bar{q}$ projector, and then using the completeness relation (4.16) for a qg pair:

$$\begin{array}{c}\rightarrow\\\rightarrow\left(\mathbf{8}\right)\rightarrow\\\leftarrow\end{array} = 2 \begin{array}{c}\rightarrow\\\rightarrow\!\!\gg\!\!\text{mmm}\!\!<\\\leftarrow\end{array} = 2 \sum_\alpha \begin{array}{c}\rightarrow\\\rightarrow\!\!\gg\!\!\left(\alpha\right)\!\!\text{m}\!\!<\\\leftarrow\end{array} , \tag{8.14}$$

where the blob α denotes the projector $\mathbb{P}_\alpha$ on the qg irrep α, given in (4.8)–(4.10) for $\alpha = \mathbf{15}, \bar{\mathbf{6}}$ and $\mathbf{3}$, respectively. Alternatively, Eq. (8.14) can be obtained by starting from (4.16) and trading the gluon for a color octet $q\bar{q}$ pair by multiplying (4.16) to the left by ⟶mm, to the right by mm⟵, and by a factor 2 to match with the $q\bar{q}$ octet projector (4.4).

Equation (8.14) provides the decomposition of the system $q(q\bar{q})_{\mathbf{8}}$,

$$\overrightarrow{\mathbf{8}} = \sum_\alpha \mathbb{P}_\alpha^{q(q\bar{q})_{\mathbf{8}}} , \tag{8.15}$$

$$\mathbf{3} \otimes \mathbf{8} = \mathbf{3}'' \oplus \bar{\mathbf{6}} \oplus \mathbf{15} , \tag{8.16}$$

where the fundamental representation appearing in the decomposition is called $\mathbf{3}''$ (to distinguish it from the irreps $\mathbf{3}'$, $\mathbf{3}_s$ and $\mathbf{3}_a$), and the $q(q\bar{q})_{\mathbf{8}}$ projectors are directly related to the qg projectors as:

$$\mathbb{P}_\alpha^{q(q\bar{q})_{\mathbf{8}}} \equiv \overrightarrow{\mathbf{8}} \left(\alpha\right) \overrightarrow{\mathbf{8}} = 2 \overrightarrow{\left(\alpha\right)} . \tag{8.17}$$

Now suppose we haven't yet found the decomposition of a qg pair into irreps. How would we proceed to find the decomposition given by Eqs. (8.15) and (8.16)? We could use the tensor method and first look for a set of independent tensors in the space $q(q\bar{q})_{\mathbf{8}}$:

$$\overrightarrow{\mathbf{8}} \left(?\right) \overrightarrow{\mathbf{8}} , \tag{8.18}$$

leading to the set of three tensors:

$$I = \overrightarrow{\mathbf{8}} ; \quad A = \overrightarrow{\mathbf{8}} \longmapsto \overrightarrow{\mathbf{8}} ; \quad B = \overrightarrow{\mathbf{8}} \bowtie \overrightarrow{\mathbf{8}} . \tag{8.19}$$

We could then establish the multiplication table between these tensors, find their characteristic equations and eigenspaces... providing a derivation very similar to that of qg irreps using the tensor method seen in Chap. 4. Another way, described below, is to derive the irreps and projectors using the index method as a guideline.

First, let's note that in terms of indices the $q(q\bar{q})_{\mathbf{8}}$ system is described by the tensor product

$$q(q\bar{q})_{\mathbf{8}} \sim q^i G_k^j = \overset{\circ \longrightarrow i}{\underset{\circ \longleftarrow k}{\overrightarrow{\mathbf{8}}}} j , \tag{8.20}$$

where G_k^j defined in (8.10) characterizes the adjoint representation.

Finding the decomposition of (8.20) using the index method goes as follows:

$$q^i G^j_k \xrightarrow[\text{one trace}]{\text{step 1}} q^l G^j_l \;:\; \text{irrep } \mathbf{3''}\,, \tag{8.21}$$

$$q^i G^j_k \xrightarrow[\text{zero trace}]{\text{step 1}} q^i G^j_k \xrightarrow{\text{step 2}} q^{\{i} G^{j\}}_k\;;\; q^{[i} G^{j]}_k \xrightarrow{\text{step 3}} T^{ij}_k\;;\; U^{ij}_k\,, \tag{8.22}$$

where the shorthand notation $\{i \ldots j\}$ (resp. $[i \ldots j]$) indicates symmetrization (resp. antisymmetrization) on indices i and j.

If we recall the discussion of the index method in Sect. 8.2, the following points should be clear:

- The projector associated to the irrep $\mathbf{3''}$ must be proportional to the tensor A defined in (8.19). Fixing the normalization one finds:

$$\mathbb{P}_{\mathbf{3''}} \equiv \underbrace{\text{8}\;\mathbf{3''}\;\text{8}} = \frac{1}{2C_F}\; \underbrace{\text{8}\;\;\text{8}}\,. \tag{8.23}$$

- Without any calculation, the projectors on the multiplets T^{ij}_k and U^{ij}_k must be:

$$\mathbb{P}_{15} = \underbrace{\text{8}\;\;\text{8}} \left(\underbrace{\text{8}} - \underbrace{\text{8}\;\mathbf{3''}\;\text{8}} \right)\,, \tag{8.24}$$

$$\mathbb{P}_{\bar{6}} = \underbrace{\text{8}\;\;\text{8}} \left(\underbrace{\text{8}} - \underbrace{\text{8}\;\mathbf{3''}\;\text{8}} \right)\,. \tag{8.25}$$

(In (8.24), the first factor ensures the appropriate symmetry in quark indices, and the irrep $\mathbf{3}_s$ in the intermediate state of this factor is removed by the second factor in brackets. In other words, the second factor ensures that the tensor T^{ij}_k is traceless.)

The above illustrates the usefulness of the index method for quickly obtaining the expressions of projectors.

In summary, by grouping first a $q\bar{q}$ pair we obtain the following decomposition of the $qq\bar{q}$ system (use (8.11) and (8.15))

$$\underbrace{\longrightarrow} = \mathbb{P}_{\mathbf{3'}} + \mathbb{P}_{\mathbf{3''}} + \mathbb{P}_{15} + \mathbb{P}_{\bar{6}}\,, \tag{8.26}$$

where the projectors are given by Eqs. (8.13) and (8.23)–(8.25).

Exercise 8.6 By expanding Eqs. (8.24) and (8.25), show that $\mathbb{P}_{15}$ and $\mathbb{P}_{\bar{6}}$ can be written alternatively as:

$$\mathbb{P}_{15} = \underbrace{\text{8}\;\;\text{8}} - \frac{1}{2(N+1)}\; \underbrace{\text{8}\;\;\text{8}}\,, \tag{8.27}$$

$$\mathbb{P}_{\bar{6}} = \overrightarrow{\underset{8}{\longrightarrow}} \blacksquare \overleftarrow{\underset{8}{\longrightarrow}} - \frac{1}{2(N-1)} \overrightarrow{\underset{8}{\longrightarrow}} \overleftarrow{\underset{8}{\longrightarrow}} . \tag{8.28}$$

Exercise 8.7 Show that $\mathbb{P}_{3'}$, $\mathbb{P}_{3''}$, $\mathbb{P}_{15}$ and $\mathbb{P}_{\bar{6}}$ project on the multiplets

$$[\mu_{3'}]^{ij}_k = \frac{1}{N} q^i q^l q_l \, \delta^j_k , \tag{8.29}$$

$$[\mu_{3''}]^{ij}_k = \frac{1}{2C_F} \left(q^l G^j_l \delta^i_k - \frac{1}{N} q^l G^i_l \delta^j_k \right) , \tag{8.30}$$

$$T^{ij}_k = \frac{1}{2} \left(q^i G^j_k + q^j G^i_k \right) - \frac{1}{2(N+1)} \left(q^l G^j_l \delta^i_k + q^l G^i_l \delta^j_k \right) , \tag{8.31}$$

$$U^{ij}_k = \frac{1}{2} \left(q^i G^j_k - q^j G^i_k \right) - \frac{1}{2(N-1)} \left(q^l G^j_l \delta^i_k - q^l G^i_l \delta^j_k \right) . \tag{8.32}$$

Explain how the multiplet (8.30) can be obtained without knowing the explicit form of the projectors (8.23)–(8.25).

8.5   Relating the Irreps Found with Different Pairings

We conclude this chapter by some exercises relating the $qq\bar{q}$ decomposition (8.26) obtained by grouping first a $q\bar{q}$ pair, and the decomposition (7.25) obtained previously by grouping first a qq pair, which we rewrite here for convenience:

$$\overrightarrow{\underset{\longleftarrow}{\overrightarrow{\longrightarrow}}} = \mathbb{P}_{3_s} + \mathbb{P}_{15} + \mathbb{P}_{3_a} + \mathbb{P}_{\bar{6}} , \tag{8.33}$$

with the projectors defined by (7.21)–(7.24).

Exercise 8.8 Since there is a single irrep **15** in the $qq\bar{q}$ decomposition (characterized by two symmetrized quark indices and one antiquark index), the traceless tensors $[\mu_{15}]^{ij}_k$ and T^{ij}_k (given respectively by (7.29) and (8.31)) arising from different pairing orders must be identical (although they look quite different because they are expressed in terms of different building blocks). Check *visually* (without calculation) that the projector (7.22) indeed coincides with (8.24). *Hint: in* (8.24), *insert the completeness relation* $\mathbf{6} \otimes \bar{\mathbf{3}} = \mathbf{3}_s \oplus \mathbf{15}$ *in the right place, then use Schur's lemma, and remember that the multiplet* μ_{15} *is traceless.*

Exercise 8.9 Check visually that the l.h.s. of (8.23) can be written as

$$\mathbb{P}_{3''} = \sum_{\alpha = 3_s, 3_a} \overrightarrow{\underset{8}{\longrightarrow}} \left(\alpha \right) \overleftarrow{\underset{8}{\longrightarrow}} , \tag{8.34}$$

and cross-check this relation by using the birdtrack expressions of $\mathbb{P}_{3_s}$, $\mathbb{P}_{3_a}$ and $\mathbb{P}_{3''}$ given in (7.21), (7.23), and in the r.h.s. of (8.23).

Exercise 8.10 In the complete $qq\bar{q}$ vector space, find a change of basis relating the irreps $\{3_s, 3_a, 15, \bar{6}\}$ to the irreps $\{3', 3'', 15, \bar{6}\}$. In other words, find an invertible map D of the $qq\bar{q}$ space such that:

$$\mathbb{P}_{3_s} = D\,\mathbb{P}_{3'}\,D^{-1}\,;\ \ \mathbb{P}_{3_a} = D\,\mathbb{P}_{3''}\,D^{-1}\,;\ \ \mathbb{P}_{15} = D\,\mathbb{P}_{15}\,D^{-1}\,;\ \ \mathbb{P}_{\bar{6}} = D\,\mathbb{P}_{\bar{6}}\,D^{-1}\,. \quad (8.35)$$

Hint: reread Sect. 3.2 carefully to first find a similarity transformation between the irreps $\mathbf{3}_s$ and $\mathbf{3}'$.

The projectors on the two equivalent irreps of the $qq\bar{q}$ system (either $\mathbf{3}'$ and $\mathbf{3}''$ in (8.26), or $\mathbf{3}_s$ and $\mathbf{3}_a$ in (8.33)) depend on the order of pairing used to decompose the system. (The sum of these projectors is of course independent of that choice, $\mathbb{P}_{3'} + \mathbb{P}_{3''} = \overrightarrow{\overset{\rightarrow}{\underset{}{}}} - \mathbb{P}_{15} - \mathbb{P}_{\bar{6}} = \mathbb{P}_{3_s} + \mathbb{P}_{3_a}$.) This is similar to the situation encountered in the case of the qqq system, where the equivalent irreps $\mathbf{8}_+$ and $\mathbf{8}_-$ in (7.15) are related to the irreps $\mathbf{8}'_+$ and $\mathbf{8}'_-$ in (7.16) (obtained with a different pairing) by a change of basis.

Chapter 9
Homework: $6 \otimes 8$ and $15 \otimes 8$

Abstract To put the techniques described in this manual into practice, we decompose into irreps the system $(qq)_6 g$ (using first the tensor method, and then the index method), and the richer system $(qg)_{15} g$ (using various tools borrowed from both the tensor method and the index method). This chapter can be considered a vacation assignment for the student addicted to birdtracks.

9.1 $6 \otimes 8$

As a first homework assignment, let us derive the irreps of the system $(qq)_6 g$ composed of a symmetric diquark state and a gluon.

9.1.1 Tensor Method

In this section we decompose $(qq)_6 g$ by applying the now-familiar "tensor method" (described in Chap. 3).

Exercise 9.1 Using the standard algorithm, show that a basis of tensors mapping $(qq)_6 g \rightarrow (qq)_6 g$ can be chosen as:

$$I = \; ; \; A = \; ; \; B = \; ; \; C = \; .$$

(9.1)

How many irreps do we expect for the system $(qq)_6 g$? Explain why the operators A, B, C are guaranteed to commute.

Exercise 9.2 Derive the multiplication table of A, B, C:

$$A^2 = C^2 = \frac{N^2+N-1}{4N} A - \frac{1}{4N} C ; \quad AC = -\frac{1}{4N} A + \frac{N^2+N-1}{4N} C ,$$

$$AB = -\frac{1}{4N} A + \frac{N-1}{4N} C ; \quad BC = \frac{N-1}{4N} A - \frac{1}{4N} C , \tag{9.2}$$

$$B^2 = \frac{1}{8} I + \frac{1}{4} B - \frac{1}{4N} (A + C) .$$

As usual, we would now like to find the characteristic equation of some of the operators A, B, C. To avoid having to do this by manipulating the multiplication table (9.2) (which can be a bit tedious), it can be convenient (Cvitanović 2008) to represent the multiplication by the operator A by a matrix $\tilde{A}$ acting in the space $\{I, A, B, C\}$ of tensors and defined by $\tilde{A} \cdot I \equiv AI = A$, $\tilde{A} \cdot A \equiv A^2$, $\tilde{A} \cdot B \equiv AB$, $\tilde{A} \cdot C \equiv AC$. (Similarly, the multiplication by B and C can be represented by some matrices $\tilde{B}$ and $\tilde{C}$.) An alternative way of presenting the multiplication table is thus to give the matrices $\tilde{A}, \tilde{B}, \tilde{C}$.

Exercise 9.3 Check that the matrices $\tilde{A}, \tilde{B}, \tilde{C}$ read (in the tensor basis $\{I, A, B, C\}$):

$$\tilde{A} = \begin{pmatrix} 0 & 0 & 0 & 0 \\ 1 & \frac{N^2+N-1}{4N} & -\frac{1}{4N} & -\frac{1}{4N} \\ 0 & 0 & 0 & 0 \\ 0 & -\frac{1}{4N} & \frac{N-1}{4N} & \frac{N^2+N-1}{4N} \end{pmatrix} , \tag{9.3}$$

$$\tilde{B} = \begin{pmatrix} 0 & 0 & \frac{1}{8} & 0 \\ 0 & -\frac{1}{4N} & -\frac{1}{4N} & \frac{N-1}{4N} \\ 1 & 0 & \frac{1}{4} & 0 \\ 0 & \frac{N-1}{4N} & -\frac{1}{4N} & -\frac{1}{4N} \end{pmatrix} ; \quad \tilde{C} = \begin{pmatrix} 0 & 0 & 0 & 0 \\ 0 & -\frac{1}{4N} & \frac{N-1}{4N} & \frac{N^2+N-1}{4N} \\ 0 & 0 & 0 & 0 \\ 1 & \frac{N^2+N-1}{4N} & -\frac{1}{4N} & -\frac{1}{4N} \end{pmatrix} . \tag{9.4}$$

The tensor A viewed as an operator of $(qq)_6 g \rightarrow (qq)_6 g$ is Hermitian and thus diagonalizable. Its minimal polynomial is therefore split with simple roots, which are the eigenvalues of A. The matrix $\tilde{A}$ acting in $\{I, A, B, C\}$ has obviously the same minimal polynomial as A. (Note that $\tilde{A}^0 \equiv \tilde{I}$, where $\tilde{I} = \text{diag}(1, 1, 1, 1)$ is the matrix representing the multiplication by I.) Thus $\tilde{A}$ is diagonalizable (even though $\tilde{A}$ is not Hermitian) and its eigenvalues coincide with those of A. The interest of the matrices $\tilde{A}, \tilde{B}, \tilde{C}$ is that they directly provide the eigenvalues of the operators A, B, C.

Exercise 9.4 Check that the eigenvalues of A are $\{a_1, a_2, a_3\} = \{\frac{1}{4} C_6, \frac{N+1}{4}, 0\}$, with $C_6 \equiv (N - 1)(N + 2)/N$. Let's denote by P_{a_i} the projector on the eigenspace of A associated with the eigenvalue a_i. Without yet knowing the explicit forms of the projectors, explain why P_{a_1} and P_{a_2} must project on two different irreps, whereas the $SU(N)$ invariant subspace $\text{img}(P_{a_3})$ needs to be further separated (see Sect. 4.2.2) to provide the two other irreps. *(Hint: pay attention to the multiplicities of the eigen-values of $\tilde{A}$.)* Finally, show that the projectors P_{a_i} are given by:

$$P_{a_1} = \frac{2}{C_6}(A+C) = \frac{4}{C_6}\;[\text{birdtrack}]\;, \tag{9.5}$$

$$P_{a_2} = \frac{2}{N+1}(A-C) = \frac{4}{N+1}\;[\text{birdtrack}]\;, \tag{9.6}$$

$$P_{a_3} = I - P_{a_1} - P_{a_2}\;. \tag{9.7}$$

Exercise 9.5 Show that the restriction of B to $\mathrm{img}(P_{a_3})$ has for minimal polynomial $t^2 - \frac{1}{4}t - \frac{1}{8} = (t - \frac{1}{2})(t + \frac{1}{4})$. Denoting its eigenvalues by $\{b_1, b_2\} = \{-\frac{1}{4}, \frac{1}{2}\}$, obtain the following expressions for the projectors on the associated eigenspaces:

$$P_{b_1} = \frac{B-b_2}{b_1-b_2}P_{a_3} = \frac{2}{3}I - \frac{4}{3}B - \frac{2}{3(N-1)}(A+C) - \frac{2}{N+1}(A-C)\,, \tag{9.8}$$

$$P_{b_2} = \frac{B-b_1}{b_2-b_1}P_{a_3} = \frac{1}{3}I + \frac{4}{3}B - \frac{4}{3(N+2)}(A+C)\,. \tag{9.9}$$

Exercise 9.6 Calculate the dimensions of the $SU(N)$ irreps characterized by the projectors $P_{a_1}, P_{a_2}, P_{b_1}, P_{b_2}$.

In summary, the system $(qq)_6 g$ decomposes into irreps as

$$6 \otimes 8 = \bar{3} \oplus 6 \oplus \overline{15} \oplus 24\,, \tag{9.10}$$

where the dimensions K_α of the $SU(N)$ irreps α are given in the following table:

Irrep α	$\bar{3}$	6	$\overline{15}$	24
K_α	$\frac{N(N-1)}{2}$	$\frac{N(N+1)}{2}$	$\frac{N^2(N^2-4)}{3}$	$\frac{(N^2-1)(N+3)N}{6}$

The associated projectors can be expressed in terms of birdtracks as follows (choosing the factorized form of the expressions (9.8) and (9.9) for $\mathbb{P}_{\overline{15}}$ and $\mathbb{P}_{24}$):

$$\mathbb{P}_{\bar{3}} = \frac{4}{N+1}\;[\text{birdtrack}]\;, \tag{9.11}$$

$$\mathbb{P}_6 = \frac{4}{C_6}\;[\text{birdtrack}]\;, \tag{9.12}$$

$$\mathbb{P}_{\overline{15}} = \left(\frac{2}{3}\;[\text{birdtrack}] - \frac{4}{3}\;[\text{birdtrack}]\right)\left(I - \mathbb{P}_{\bar{3}} - \mathbb{P}_6\right)\,, \tag{9.13}$$

$$\mathbb{P}_{24} = \left(\frac{1}{3}\;[\text{birdtrack}] + \frac{4}{3}\;[\text{birdtrack}]\right)\left(I - \mathbb{P}_{\bar{3}} - \mathbb{P}_6\right)\,, \tag{9.14}$$

and obviously satisfy the completeness relation:

$$\mathbb{P}_{\bar{3}} + \mathbb{P}_6 + \mathbb{P}_{\overline{15}} + \mathbb{P}_{24} \;=\; \text{⟨diagram⟩} \;. \tag{9.15}$$

Let us emphasize that the SU(N) irreps $\bar{3}$, **6**, $\overline{15}$ of the $(qq)_6 g$ system and the SU(N) irreps **3**, $\bar{6}$, **15** of a qg pair (see Chap. 4, Eqs. (4.15) and (4.12)) are complex conjugate to each other for $N = 3$, but have nothing in common for $N > 3$, as can be seen by comparing their dimensions as a function of N.

9.1.2 Index Method

In Chap. 8, we learnt how to decompose the product of two irreps by a mere observation of the quark and antiquark indices carried by the irreps, using the "index method". As a second homework, we obtain the irreps and associated projectors of $(qq)_6 g$ with this method, allowing a shorter derivation than with the tensor method used in the previous section.

First, in order to single out quark and antiquark indices, we trade the gluon for a $q\bar{q}$ pair in the SU(N) adjoint representation:

$$\text{⟨diagram⟩} \;=\; 2\;\text{⟨diagram⟩} \;. \tag{9.16}$$

Looking at the intermediate state and denoting the quark and antiquark indices by $ijkl$ (from bottom to top), we see that $(qq)_6 g \sim \mu_S^{ij} G_l^k$, with μ_S^{ij} and G_l^k the multiplets characterizing a symmetric diquark state (see Chap. 3) and a SU(N) gluon (see Exercise 8.2 and Eq. (8.10)).

According to the index method, the product $\mu_S^{ij} G_l^k$ decomposes as:

$$\mu_S^{ij} G_l^k \xrightarrow[\text{one trace}]{\text{step 1}} \mu_S^{mj} G_m^k \xrightarrow{\text{step 2}} \mu_S^{m\{j} G_m^{k\}} \;;\; \mu_S^{m[j} G_m^{k]} \;:\; \text{irreps } \mathbf{6} \text{ and } \bar{\mathbf{3}} \;, \tag{9.17}$$

$$\mu_S^{ij} G_l^k \xrightarrow[\text{zero trace}]{\text{step 1}} \mu_S^{ij} G_l^k \xrightarrow{\text{step 2}} \mu_S^{(ij} G_l^{k)}{}_{10} \;;\; \mu_S^{(ij} G_l^{k)}{}_8 \xrightarrow{\text{step 3}} [\mu']_l^{ijk} \;;\; [\mu'']_l^{ijk} \tag{9.18}$$

Let's first consider the irreps arising from taking one trace in the product $\mu_S^{ij} G_l^k$, given by the line (9.17) of the procedure. Observing the intermediate state of (9.16), there is clearly only one way to contract the antiquark index l with a quark index (due to the symmetry in ij of μ_S^{ij}, and the fact that G_l^k is traceless). This means that the tensors needed to construct the projectors on those irreps are of the form:

$$\text{⟨diagram⟩} \;. \tag{9.19}$$

There are now two ways of connecting the two quark lines entering the grey blob from the left to those leaving from the right. We recover the tensors A and C identified in Sect. 9.1.1 using the tensor method. But the index method tells us that if we trade these two ways of connecting lines for the two possible symmetries in the quark indices jk (step 2 of line (9.17)), we obtain directly the irreps. In other words, the projectors on the two irreps must be of the form:

$$\mathbb{P}_6 = c_1 \; \text{⟶⊞⟶⊞⟶} \; ; \quad \mathbb{P}_{\bar{3}} = c_2 \; \text{⟶⊞⟶■⟶} \; . \tag{9.20}$$

Looking at the intermediate states and using Schur's lemma, it is obvious that the irreps characterized by the projectors (9.20) are equivalent to the diquark irreps **6** and $\bar{\mathbf{3}}$, hence the names given to the projectors.

Exercise 9.7 Find the normalization factors c_1 and c_2 so that $\mathbb{P}_6$ and $\mathbb{P}_{\bar{3}}$ are indeed projectors. Explain pictorially why $c_1 = 4/C_6$, where C_6 is the Casimir charge of the diquark irrep **6**, without even knowing C_6 as a function of N. (If needed, the Casimirs of the diquark irreps were evaluated in Exercise 3.6.)

Let's now consider the irreps of $(qq)_6 g$ arising from taking *zero* trace in $\mu_S^{ij} G_l^k$, corresponding to the procedure (9.18). In step 2 of this procedure, we have to find all possible symmetries in the quark indices ijk, knowing that ij are symmetrized in μ_S^{ij}. This amounts to finding the irreps of $\mu_S^{ij} q^k \sim \mathbf{6} \otimes \mathbf{3}$, which was already done in Sect. 7.1.1, see (7.7) and the projectors $\mathbb{P}_{10}$ and $\mathbb{P}_{8_+}$ given by (7.5) and (7.6), whose pictorial form specifies the symmetry in ijk. In the tensors resulting from step 2, $\mu_S^{(ij} G_l^{k)10}$ and $\mu_S^{(ij} G_l^{k)8}$, the symmetry in ijk is thus indicated by the subscripts **10** and **8**, respectively, and is realized pictorially by projecting the qqq system in the intermediate state of (9.16) on the qqq irrep **10** or $\mathbf{8}_+$. In step 3, these two tensors are made traceless (to provide the multiplets μ' and μ'') by simply removing from the full $qqq\bar{q}$ intermediate state the irreps **6** and $\bar{\mathbf{3}}$ characterized by a larger number of traces (similarly to what was done for the system $q(q\bar{q})_8$ in Sect. 8.4, see (8.24) and (8.25)). With a little practice, one can thus directly infer from the procedure (9.18) the projectors on the multiplets μ' and μ'':

$$\mathbb{P}' = 2 \; \text{⟶⊞⟶} \boxed{10} \text{⟶⊞⟶} \; \left(I - \mathbb{P}_{\bar{3}} - \mathbb{P}_6 \right) , \tag{9.21}$$

$$\mathbb{P}'' = 2 \; \text{⟶⊞⟶} \boxed{8_+} \text{⟶⊞⟶} \; \left(I - \mathbb{P}_{\bar{3}} - \mathbb{P}_6 \right) , \tag{9.22}$$

where the blobs labelled **10** and $\mathbf{8}_+$ stand for the qqq projectors (7.5) and (7.6).

Exercise 9.8 Using (9.21) and (9.22), verify explicitly that $\mathbb{P}'$ and $\mathbb{P}''$ are projectors, orthogonal to each other and to $\mathbb{P}_{\bar{3}}$ and $\mathbb{P}_6$.

This shows that $\mathbb{P}_{\bar{3}}$, $\mathbb{P}_6$, $\mathbb{P}'$ and $\mathbb{P}''$ are the projectors on the four irreps of the $(qq)_6 g$ system we are looking for. Thus, the set of projectors $\{\mathbb{P}', \mathbb{P}''\}$ must coincide with the set $\{\mathbb{P}_{\overline{15}}, \mathbb{P}_{24}\}$ found using the tensor method, see (9.13) and (9.14).

Exercise 9.9 For $N = 3$, explain why $\mathbb{P}''$ projects on $\overline{\mathbf{15}}$ (and not $\mathbf{24}$), by simply observing the pictorial expression (9.22). As a cross-check, verify that (9.13) and (9.22) coincide for all N.

9.2 $\mathbf{15} \otimes \mathbf{8}$

In order to test your expertise in the birdtrack technique, as a final homework assignment we consider the system $(qg)_{15} g$, made of a quark-gluon pair in the irrep $\mathbf{15}$ (of associated projector defined by (4.8) or equivalently (4.17)) and a gluon. This system is richer than the systems considered so far, and the derivation of its irreps is facilitated by the combined use of tensor and index methods.

9.2.1 Decomposition into SU(N) Irreps

We start by representing the identity of the $(qg)_{15} g$ vector space as

$$
I \;=\; \text{(15)} \;=\; 4 \; \text{(15)} \quad \text{(15)} \;, \tag{9.23}
$$

where we used (4.17).

Exercise 9.10 In the intermediate state of (9.23), let us denote the quark and anti-quark indices (from bottom to top) by $ijklm$. By applying the index method to the product $[\mu_{15}]_k^{ij} G_m^l$, show that the system $(qg)_{15} g$ decomposes into a sum of SU(N) irreps as

$$
\mathbf{15} \otimes \mathbf{8} \;=\; \mathbf{3} \oplus \bar{\mathbf{6}} \oplus \mathbf{15}_a \oplus \mathbf{15}_b \oplus \mathbf{15}' \oplus \overline{\mathbf{24}} \oplus \mathbf{42} \oplus \mathbf{0} \;, \tag{9.24}
$$

and calculate the dimensions of these irreps as a function of N. Explain briefly why for any N, the irreps $\mathbf{15}_a$, $\mathbf{15}_b$ and the quark-gluon pair irrep $\mathbf{15}$ are equivalent, whereas the irrep $\mathbf{15}'$ is *not* equivalent to these three irreps.

Let's give a hint to obtain the dimension of one of the irreps, e.g. the irrep R arising from taking zero trace in $[\mu_{15}]_k^{ij} G_m^l$, and being totally symmetric in quark indices and antisymmetric in antiquark indices. This irrep is associated to the tensor $[\mu_{15}]_{[k}^{(ij} G_{m]}^{l)}$, and corresponds to an intermediate state in (9.23) made of a qqq system in the irrep $\mathbf{10}$ and a $\bar{q}\bar{q}$ pair in the irrep $\mathbf{3}$. The dimension of R is thus given by the dimension of the space $(qqq)_{10}(\bar{q}\bar{q})_3 \sim \mathbf{10} \otimes \mathbf{3}$ (which is easy to find), minus the number of constraints needed to make $[\mu_{15}]_{[k}^{\{ij}G_{m]}^{l\}}$ traceless. The latter number equals the sum of dimensions of the irreps found in the decomposition of $\mathbf{10} \otimes \mathbf{3}$ and corresponding to a non-zero number of traces. Thus, calculating the dimension of R requires finding the irreps of $(qqq)_{\mathbf{10}}(\bar{q}\bar{q})_{\mathbf{3}}$ as a preliminary exercise (which can be quickly done using the index method). You will easily find that the irrep of the system $(qg)_{\mathbf{15}}g$ under consideration is $R = \mathbf{15}'$ (and incidentally observe that $\mathbf{10} \otimes \mathbf{3} = \mathbf{15} \oplus \mathbf{15}'$).

Exercise 9.10 illustrates that the dimensions of the irreps arising from different numbers of traces have a different scaling in N when N increases. (This can also be verified for the other parton systems studied with the index method, see Chap. 8 and Sect. 9.1.2.) The irreps obtained from a larger number of traces have a smaller dimension at sufficiently large N. In the following we derive the Hermitian projectors associated to the irreps of the decomposition (9.24), for the lower-dimensional irreps $\mathbf{3}$, $\bar{\mathbf{6}}$, $\mathbf{15}_a$, $\mathbf{15}_b$ in Sect. 9.2.2, and for the higher-dimensional irreps $\mathbf{15}'$, $\overline{\mathbf{24}}$, $\mathbf{42}$, $\mathbf{0}$ in Sect. 9.2.3.

9.2.2 Projectors on Lower-Dimensional Irreps

In the index method, the irrep $\mathbf{3}$ appearing in (9.24) results from taking *two* traces in the product $[\mu_{15}]_k^{ij}G_m^l$.

Exercise 9.11 Explain why the projector $\mathbb{P}_3$ on the irrep $\mathbf{3}$ must be of the form

$$\mathbb{P}_3 = c_3 \quad \begin{array}{c}\text{\scriptsize 15}\end{array} \quad \begin{array}{c}\text{\scriptsize 15}\end{array} \quad , \qquad (9.25)$$

and evaluate the normalization coefficient c_3.

As for the irreps $\bar{\mathbf{6}}$, $\mathbf{15}_a$ and $\mathbf{15}_b$ of the decomposition (9.24), they arise from $[\mu_{15}]_k^{ij}G_m^l$ by taking *one* contraction between a quark index and an antiquark index. In order to find the projectors associated with these irreps, let us first observe that there are five independent tensors mapping $(qg)_{\mathbf{15}}g \rightarrow (qg)_{\mathbf{15}}g$ with an intermediate state satisfying this criterion, which can be chosen as:

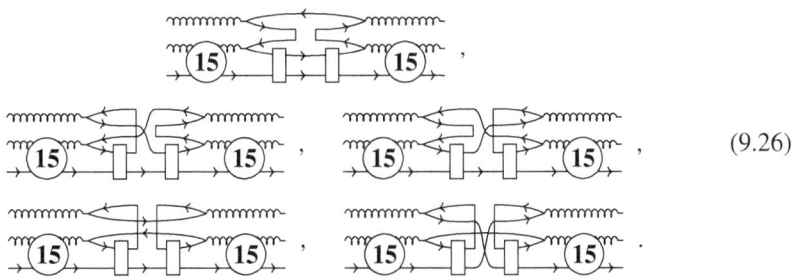

$$(9.26)$$

Exercise 9.12 Why five and not just three?

Exercise 9.13 Show that the first three tensors of (9.26) can be redrawn as

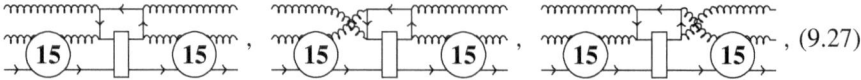

$$(9.27)$$

where the diquark symmetrizer is optional and could be replaced by $\xrightarrow{\hspace{0.4cm}}$ if needed, and that the last two tensors of (9.26) can be traded for

$$(9.28)$$

where the diquark symmetrizer or antisymmetrizer is *not* optional. Explain why, of the five tensors (9.27) and (9.28), the last must be the tensor needed to construct the projector $\mathbb{P}_{\bar{6}}$ on the irrep $\bar{\mathbf{6}}$ of the decomposition (9.24), and the other four those needed to construct the projectors $P_a \equiv \mathbb{P}_{15_a}$, $P_b \equiv \mathbb{P}_{15_b}$ on the irreps $\mathbf{15}_a$ and $\mathbf{15}_b$, as well as the transition operators T and $T^\dagger$ for $\mathbf{15}_a \rightarrow \mathbf{15}_b$ and $\mathbf{15}_b \rightarrow \mathbf{15}_a$, respectively.

Since any intermediate state of the operators P_a, P_b, T, $T^\dagger$ (resp. $\mathbb{P}_{\bar{6}}$) must belong to the irrep $\mathbf{15}$ (resp. $\bar{\mathbf{6}}$), we can readily remove the unnecessary irreps from the $qq\bar{q}$ intermediate state of the tensors (9.27) and (9.28). For the first four tensors, we decompose the $(qq)_6\bar{q}$ intermediate state as $\mathbf{6} \otimes \bar{\mathbf{3}} = \mathbf{3}_s \oplus \mathbf{15}$ (see Exercise 7.5), but keep only the contribution from the irrep $\mathbf{15}$. For the last tensor of (9.28), we decompose $(qq)_{\bar{3}}\bar{q}$ as $\bar{\mathbf{3}} \otimes \bar{\mathbf{3}} = \mathbf{3}_a \oplus \bar{\mathbf{6}}$ (see Exercise 7.6), and keep only the $\bar{\mathbf{6}}$. We therefore introduce the tensors

$$A = \quad (9.29)$$

$$B_{11} = \qquad , \quad B_{22} = $$

$$B_{12} = \qquad , \quad B_{21} = \qquad , \quad (9.30)$$

where the large discs denoted by $\mathbf{15}$ and $\bar{\mathbf{6}}$ represent the already known $qq\bar{q}$ projectors (7.22) and (7.24), respectively, so that the tensors A and B_{ij} (i, $j = 1$, 2) are explicitly given in terms of birdtracks.

It should now be clear that A is proportional to the projector on the irrep $\bar{\mathbf{6}}$ of $(qg)_{15}g$, and that some linear combinations of the tensors B_{ij} yield the operators P_a, P_b, T, $T^\dagger$. In other words, the tensors B_{ij} and the tensors P_a, P_b, T, $T^\dagger$ span the same space, denoted by $\mathcal{T}_{15}$ in the following.

Exercise 9.14 From Schur's lemma, we must have $A^2 = cA$. Find the coefficient c from a relatively simple birdtrack calculation. (One may need the expression (4.8) of the projector on the qg irrep **15**.) Deduce the projector $\mathbb{P}_{\bar{6}}$ on the irrep $\bar{\mathbf{6}}$ of (9.24).

We now want to determine which linear combinations of the B_{ij}'s correspond to the projectors P_a and P_b on the irreps $\mathbf{15}_a$ and $\mathbf{15}_b$. In fact, as these irreps are equivalent, P_a and P_b are not uniquely determined (as we are now used to). To deal with this situation, we first establish the multiplication table of the B_{ij} tensors. (Thanks to the elimination of unwanted irreps to obtain the tensors (9.29) and (9.30), the B_{ij}'s form a closed set under multiplication).

Exercise 9.15 Using Schur's lemma, show that the multiplication of the B_{ij}'s can be written as :

$$B_{ij} B_{kl} = n_{jk} B_{il} , \qquad (9.31)$$

involving four numbers n_{jk} ($j, k = 1, 2$). Give the birdtrack expressions of n_{jk}, and show that $n_{12} = n_{21}$. (The explicit calculation of n_{11}, n_{22}, n_{12} as a function of N is postponed to Exercise 9.20.)

Exercise 9.16 Similarly to what was done in Sect. 9.1.1 (see Exercise 9.3), in the space $\mathcal{T}_{15}$ we can represent the multiplication by B_{ij} by a 4×4 matrix $\tilde{B}_{ij}$. Express the multiplication table (9.31) alternatively by writing the matrices $\tilde{B}_{11}$, $\tilde{B}_{22}$, $\tilde{B}_{12}$, $\tilde{B}_{21}$ (in terms of n_{11}, n_{22}, n_{12}) in the basis $\{B_{11}, B_{22}, B_{12}, B_{21}\}$ of $\mathcal{T}_{15}$.

Exercise 9.17 In the space $\mathcal{T}_{15}$, show that $P_a + P_b$ is the identity operator and that $(T - T^\dagger)^2 \propto P_a + P_b$. *Hint: study the action of $P_a + P_b$ and $(T - T^\dagger)^2$ in the basis $\{P_a, P_b, T, T^\dagger\}$ of $\mathcal{T}_{15}$, and remember the general expression of a transition operator.*

Exercise 9.18 Using the results of the previous two exercises, show that:

$$P_a + P_b \equiv \mathbb{P}_{15_a} + \mathbb{P}_{15_b} = \frac{n_{22} B_{11} + n_{11} B_{22} - n_{12} \left(B_{12} + B_{21}\right)}{n_{11} n_{22} - n_{12}^2} . \qquad (9.32)$$

Hint: in $\mathcal{T}_{15}$, only one anti-Hermitian operator can be built (up to a global factor).

Exercise 9.19 The previous exercises show that $P_a + P_b$ is uniquely determined. To check that P_a and P_b are not uniquely determined separately, propose two different choices for the set of projectors $\{P_a, P_b\}$. For each choice, verify that P_a and P_b are orthogonal to each other and each have a rank equal to the dimension of the irrep **15**.

Exercise 9.20 Finally, let us derive n_{11}, n_{22}, n_{12} as a function of N from their birdtrack expressions found in Exercise 9.15. With a little care and ingenuity, each coefficient can be calculated by hand on half a page. The results given below contain two misprints. Please correct them!

$$n_{11} = \frac{N^3 + N^2 - 3N - \sqrt{5}}{4N(N+1)} ; \; n_{22} = \frac{N^3 + 2N^2 - 2N - 2}{\pi N(N+1)} ; \; n_{12} = -\frac{2N+1}{4N(N+1)} .$$

$$(9.33)$$

9.2.3 Projectors on Higher-Dimensional Irreps

In this last section, we derive the projectors on the irreps $\mathbf{15'}, \overline{\mathbf{24}}, \mathbf{42}, \mathbf{0}$, arising from the product $[\mu_{15}]^{ij}_k \, G^l_m$ without performing any contraction between upper and lower indices. Compared with the derivation of projectors on lower-dimensional irreps in the previous section, this is the easy part. If you have digested the index method, you can directly state the results, given in Eqs. (9.36)–(9.39) below, without any justification. If this is not the case, you can read the following explanations to refine your intuition a little more before closing this manual.

After using (9.23) to trade the gluons of the system $(qg)_{15}g$ for octet $q\bar{q}$ pairs, the index method tells us that the irreps with zero trace we are looking for are characterized by their symmetries in three quark indices and two antiquark indices. Let's focus on the irrep $\mathbf{42}$ arising from the tensor $[\mu_{15}]^{\{ij}_{\{k} \, G^{l\}}_{m\}}$ (obtained after step 2), totally symmetric in quark and separately antiquark indices. Remember that this tensor must be made traceless (step 3), to yield the multiplet defining the irrep. One can directly write the associated projector as:

$$\mathbb{P}_{42} \;=\; 4 \;\text{(15)} \underset{}{\Longrightarrow} \left(\;\parallel\; - \;\overset{3}{\bigcirc}\; - \;\overset{15}{\bigcirc}\; \right) \underset{}{\Longleftarrow} \text{(15)} \;,$$
(9.34)

where the factor in brackets guarantees that the $qqq\bar{q}\bar{q}$ intermediate state is itself in the irrep $\mathbf{42}$ (as it should from Schur's lemma): the first term ensures the required symmetry in quark and antiquark indices, and the subtraction terms ensure the tracelessness of the associated multiplet by removing the irreps $\mathbf{3}$ and $\mathbf{15}$ of the system $(qqq)_{10}(\bar{q}\bar{q})_{\bar{6}}$. (The large discs in (9.34) denote the projectors on those irreps.)

Indeed, by applying the index method to $q^{\{i}q^jq^{l\}}q_{\{k}q_{m\}} \equiv \psi^{ijl}\varphi_{km}$, we find that $(qqq)_{10}(\bar{q}\bar{q})_{\bar{6}}$ decomposes as

$$\mathbf{10} \otimes \overline{\mathbf{6}} = \mathbf{3} \oplus \mathbf{15} \oplus \mathbf{42} \,,$$
(9.35)

where $\mathbf{15}$ and $\mathbf{3}$ arise from taking respectively one and two traces in $\psi^{ijl}\varphi_{km}$. Thus, in (9.34) the factor in brackets is nothing more than the projector on the irrep $\mathbf{42}$ of the $qqq\bar{q}\bar{q}$ intermediate state, obviously characterized by the same number of traces (zero) and the same symmetries as the irrep $\mathbf{42}$ of the system $(qg)_{15}g$.

In order to obtain the projector (9.34) in explicit form, we could of course calculate the projectors on the irreps $\mathbf{3}$ and $\mathbf{15}$ of the system $(qqq)_{10}(\bar{q}\bar{q})_{\bar{6}}$, but the following exercise shows that this is not necessary.

Exercise 9.21 Using (9.35) and Schur's lemma, show that (9.34) can be rewritten as follows:

$$\mathbb{P}_{42} \;=\; 4 \;\text{(15)} \underset{}{\Longrightarrow} \parallel \underset{}{\Longleftarrow} \text{(15)} \left(I - \mathbb{P}_3 - \mathbb{P}_{\bar{6}} - \mathbb{P}_{15_a} - \mathbb{P}_{15_b} \right), \quad (9.36)$$

where I is defined by (9.23), the projectors $\mathbb{P}_3$ and $\mathbb{P}_{\bar{6}}$ were found in Exercises 9.11 and 9.14, and $\mathbb{P}_{15_a} + \mathbb{P}_{15_b}$ is given by (9.32).

The expression (9.36) is a fully explicit form of the projector $\mathbb{P}_{42}$ (acting on the system $(qg)_{15}g$) in terms of birdtracks. The above discussion and Exercise 9.21 illustrate that in order to derive the projector on an irrep R characterized by a given number of traces (and some symmetries), the subtraction removing the irreps characterized by a larger number of traces (or, equivalently, making the multiplet associated to R traceless), can be performed at the level of any intermediate state of the parton system (including the initial state, or final state as in Eq (9.36)). In practice, the intermediate state is chosen so that the subtraction involves already known projectors.

Following the same procedure, we easily obtain the projectors on the remaining irreps $\mathbf{15'}$, $\overline{\mathbf{24}}$, and $\mathbf{0}$:

$$\mathbb{P}_{15'} = 4 \; \boxed{\text{15}} \cdots \boxed{\text{15}} \left(I - \mathbb{P}_3 - \mathbb{P}_{\bar{6}} - \mathbb{P}_{15_a} - \mathbb{P}_{15_b} \right), \quad (9.37)$$

$$\mathbb{P}_{\overline{24}} = 4 \; \boxed{\text{15}} \boxed{\text{8}} \boxed{\text{15}} \left(I - \mathbb{P}_3 - \mathbb{P}_{\bar{6}} - \mathbb{P}_{15_a} - \mathbb{P}_{15_b} \right), \quad (9.38)$$

$$\mathbb{P}_{0} = 4 \; \boxed{\text{15}} \boxed{\text{8}} \boxed{\text{15}} \left(I - \mathbb{P}_3 - \mathbb{P}_{\bar{6}} - \mathbb{P}_{15_a} - \mathbb{P}_{15_b} \right), \quad (9.39)$$

where the disc denoted by **8** is the qqq projector (7.6) (of mixed symmetry in the three quark indices).

Exercise 9.22 Show to your neighbour in less than two minutes, using (9.36)–(9.39) and remembering the discussion around (9.34) and (9.35), that $\mathbb{P}_{42}$, $\mathbb{P}_{15'}$, $\mathbb{P}_{\overline{24}}$, $\mathbb{P}_0$ are indeed projectors, orthogonal to each other and to $\mathbb{P}_3$, $\mathbb{P}_{\bar{6}}$, $\mathbb{P}_{15_a}$, $\mathbb{P}_{15_b}$, and satisfy the completeness relation $I = \mathbb{P}_3 + \mathbb{P}_{\bar{6}} + \mathbb{P}_{15_a} + \mathbb{P}_{15_b} + \mathbb{P}_{15'} + \mathbb{P}_{\overline{24}} + \mathbb{P}_{42} + \mathbb{P}_0$.

Now that we've come to the end of our introduction to color in QCD, it's worth taking a look at the three legos (1.1), of which all the concepts and results presented in this manual are direct consequences. In fact, as the second and third legos are entirely determined by the first (respectively by a change of sign and via Eq. (3.8)), we can even say that all the information contained in this manual is stored somewhere in the first lego (the SU(N) generator in the fundamental representation).

Exercise 9.23 Meditate on that! *Hint: lie on the floor, arms wide apart, eyes closed, and visualize a blue sky full of birdtracks.*

References

Alcock-Zeilinger, J., Weigert, H.: Simplification rules for Birdtrack operators. J. Math. Phys. **58**(5), 051701 (2017). https://doi.org/10.1063/1.4983477, arXiv:1610.08801 [math-ph]. (Compact Hermitian young projection operators. J. Math. Phys. **58**(5), 051702 (2017). https://doi.org/10.1063/1.4983478, arXiv:1610.10088 [math-ph], and Transition operators. J. Math. Phys. **58**(5), 051703 (2017). https://doi.org/10.1063/1.4983479, arXiv:1610.08802 [math-ph])

Blok, B., Dokshitzer, Y., Frankfurt, L., Strikman, M.: Phys. Rev. D **83**, 071501 (2011). https://doi.org/10.1103/PhysRevD.83.071501. arXiv:1009.2714 [hep-ph]

Collins, J.: Foundations of perturbative QCD. Camb. Monogr. Part. Phys. Nucl. Phys. Cosmol. **32**, 1–624 (2011). ISBN 978-1-00-940184-5, https://doi.org/10.1017/9781009401845. (Cambridge University Press, 2013)

Cougoulic, F., Peigné, S.: Nuclear $p_\perp$-broadening of an energetic parton pair. JHEP **05**, 203 (2018). https://doi.org/10.1007/JHEP05(2018)203. arXiv:1712.01953 [hep-ph]

Cvitanović, P.: Group Theory: Birdtracks, Lie's and Exceptional Groups. Princeton University Press, Princeton, NJ (2008). https://www.birdtracks.eu

Dokshitzer, Yu.L.: Perturbative QCD (and beyond). In: Strong Interactions Study Days Kloster Banz, Germany, Oct 10–12 (1995). https://doi.org/10.1007/BFb0105858. (Lecture Notes Physics, vol. 496, pp. 87 (1997))

Dokshitzer, Y.L., Marchesini, G.: Soft gluons at large angles in hadron collisions. JHEP **01**, 007 (2006). arXiv:hep-ph/0509078

Georgi, H.: Lie Algebras In Particle Physics: From Isospin To Unified Theories. CRC Press, Taylor & Francis (2000). https://doi.org/10.1201/9780429499210

Hamermesh, M.: Group Theory and its Application to Physical Problems. Addison-Wesley, Reading, MA (1962)

Keppeler, S., Sjodahl, M.: Orthogonal multiplet bases in SU(Nc) color space. JHEP **09**, 124 (2012). https://doi.org/10.1063/1.4865177. arXiv:1207.0609 [hep-ph]. (and Hermitian Young Operators, J. Math. Phys. 55 (2014), 021702, arXiv:1307.6147 [math-ph]. https://doi.org/10.1007/JHEP09(2012)124)

Keppeler, S.: Birdtracks for SU(N). In: QCD Master Class 2017, Saint-Jacut-de-la-Mer, France, June 18–24 (2017). https://doi.org/10.21468/SciPostPhysLectNotes.3. arXiv:1707.07280. (SciPost Physics Lecture Notes, vol. 3, 2018)

Kuipers, J., Ueda, T., Vermaseren, J.A.M., Vollinga, J.: FORM version 4.0. Comput. Phys. Commun. **184**, 1453–1467 (2013). arXiv:1203.6543

MacFarlane, A.J., Sudbery, A., Weisz, P.H.: On Gell-Mann's lambda-matrices, d- and f-tensors, octets, and parametrizations of SU(3). Commun. Math. Phys. **11**, 77–90 (1968). https://doi.org/10.1007/BF01654302

Peigné, S.: Introduction to Color in QCD: Initiation to the Birdtrack Pictorial Technique (2023). arXiv:2302.07574 [hep-ph]

© The Editor(s) (if applicable) and The Author(s), under exclusive license to Springer Nature Switzerland AG 2024
S. Peigné, *Color in QCD*, SpringerBriefs in Physics,
https://doi.org/10.1007/978-3-031-53681-6